Merbeings:
The True Story of Mermaids, Mermen, and Lizardfolk

By Mark A. Hall, Loren Coleman, and David Goudsward

Anomalist Books
Charlottesville, Virginia

CONTENTS

Dedicated to Shelia Hall, Jenny Coleman, and Heather Bernard

Preface: From Legend to Reality

SOME PEOPLE INSIST THAT CRYPTOZOOLOGY pursues only phantoms. However, time is the ally of the cryptozoologist. The march of progress churns up evidence that supports our findings rather than discourages the effort. Newly discovered small animals turn up all the time, like the rodent called *Kha-nyou* (Laotian Rock Rat). The Ivory-billed Woodpecker is thought to have come back from "extinction" in Arkansas. The latest blockbuster for scientists was the realization that "Little People"—now called *Homo floresiensis*—had been living in Indonesia until relatively recently. The find prompted the senior editor at *Nature* magazine to remark, "Now, cryptozoology, the study of such fabulous creatures, can come in from the cold."[1]

The first step in cryptozoology is the same in discussing any subject burdened with a body of folklore attached to it. This book will ask you to put aside your preconceptions of mermen and mermaids, an archetype that still exists. In 1967, a mermaid was spotted on the rocks at the west entrance to Active Pass, heavily trafficked by ferry routes out of Vancouver and Vancouver Island.

"Several witnesses said the mermaid had a large coho salmon, and one witness swore she had taken a bite out of it," reads an article from *The Daily Colonist*. "Long, silver-blonde hair and topless condition were generally agreed upon."[2]

There were multiple witnesses on the ferries, photographs that made the front page, and even aerial footage. The account occasionally appears in cryptozoology books with no further details. The "dimpled mermaid" from the article quickly became a local celebrity. Two days later, a reward for the mermaid was offered by the aquarium Pacific Undersea Gardens. This glass-enclosed underwater tourist attraction was rapidly becoming a sinking ship.[3] Was it a coincidence that the mermaid conveniently disappeared before anyone could capture her? She reappeared a week later and then vanished, no longer newsworthy.[4] Her disappearance as interest waned was not unexpected. Also not surprising was the wake of coverage and free publicity for the aquarium she left behind.

This sort of dubious report is precisely what we need to avoid. It supports a mental image accumulated over years of exposure to fairy tales, children's books, commercialism, fiction, and motion picture make-believe. While some of this mass-market media might come close to reality, the majority represents a misleading barrier to overcome. If we conjure up fictional images from the start, we are thwarted before we have begun.

In the following pages, we will explore the world beyond the familiar, routine events we are so familiar with. First, we will establish new images of these water creatures based on the best sources of knowledge available to us. We'll reveal the records of actual encounters with the extraordinary and those finding their traces. And find some concrete sources, in the form of fossils, that will show us there were precursors to the merbeings that some people have been lucky enough to see in living form.

Put aside notions of these beings as half-man and half-fish. They are no such thing. Put aside your fear of unfamiliar aquatic creatures, some reptilian in appearance, that emerge from swamps, rivers, and oceans to walk among humanity. The more you know and understand these creatures, the less you'll fear their differences. They are primates like ourselves, but are highly specialized in form and abilities. They have an ancient ancestry and a story of survival that we are only beginning to uncover in our human pursuit of understanding Earth's past. We will dive into all of this.

The very idea of aquatic apes is controversial. How could something so universally derided be taken seriously? The answer is that there is a long history of such creatures. They have been reported as real beings for centuries, not just described in legends and mythology; there is also a large body of such folklore. And what is folklore but the dustbin into which things not easily cataloged have been swept? The reports run right up to modern times, even though sightings often aren't taken seriously.

There is even a basis for such aquatic beings in the fossil records of the primates. The water ape is only one category in an extensive collection of remarkable mysterious primate creatures. They all have a basis in the fossil record. Paleoanthropologists are reluctant to acknowledge these connections to extant primates, relegating them to languish in obscurity. Hominology and cryptozoology believe such relationships are possible and worthy of discussion. This book is one in a series that describes these living fossils.[5] The people sorting the fossils fear the scorn of their conservative and timid colleagues if they examine the possible survival of some of these creatures. This is a natural course of events for controversial ideas to be rejected at first.[6] Some things are accepted more quickly than others.

The discovery of *Homo floresiensis*, announced in October 2004 (as well as *Homo luzonensis*, 2010, and more specimens in 2019), shocked many modern scientists. Only those of us who paid heed to the discussions of proto-pygmies and Little People that had been ongoing for half a century were not surprised. We have supported and participated in those discussions about the existence of Little People. There are physical descriptions from modern sightings supported by widespread folklore. Diminutive footprints have been discovered and evidence of such creatures can be found in the ancient fossil records of primates. The existence of the water apes can be posited on the same basis.

We need not be surprised at the presence of these primates in the natural world if we pay attention to the traces they have left for us to find, just as *Homo floresiensis* has done. The fossil finds discovered over the last several millions of years will give us no choice but to recognize the existence of more unknown species from the period when primates were proliferating. *Homo sapiens* were particularly successful, but others survived by avoiding humans and finding their own places in nature to thrive. And the remarkable creature, the water ape, came about in this period of primate abundance.

The theory is straightforward. Sixteen million years ago, there was a great flourishing of apes on the planet. Their diversity allowed them to spread across the world's habitats and establish themselves in enduring ways, and their descendants still exist. They might be around in fewer numbers than they once were, but they have persisted right up to our time. Those closest to humans have evolved and proliferated in the last few million years. For this reason, we believe we have some very close relatives living in the wildernesses of the world. The water apes are one of the more distant relatives.

They are one of the apes that ventured into the aquatic habitats of the planet and succeeded there. The competition was too great on dry land. We know this because we have many examples of ancient fossil primates with similar capabilities. And their descendants have been seen to this day, accounting for the many reports of Yeti, Abominable Snowman, and Bigfoot creatures populating most of the major continents.

The ancestor of these water apes retreated, first into the swamps and then deeper bodies of water where there was less competition. They evolved a particular way of living suited to freshwater and marine environments over millions of years. As the contention for space and resources continued around them on land, the water apes made their way around the planet's aquatic realms. But, like their land relatives, they still exist.

They have earned a place in our history, art, and mind, while remaining an enigma to traditional science. That is about to change. In a comprehensive search for fossil remains of the past, these peculiar denizens of the water world will turn up in greater numbers. They will be found to have a wide distribution in both early and recent fossil deposits.

As their fossils are discovered, we will realize their presence is part of our present. If we choose to do so, we can embark now on a journey of learning about water apes as part of our present-day world. This, too, will be discussed in this book.

Different places and cultures have freely applied many names to the water apes under discussion. Local and regional names will often be unfamiliar, especially compared to classical references such as the Nereids and the Hamadryads. Overarching categories such as merfolk and merbeings will be used when grouping the more familiar stories of mermen and mermaids. The title of this book is a nod to popular culture. For example, when the appearance of a water ape made headlines in 1988, the media dubbed it a "Lizardman." He has been remembered by this name for his colorful place in modern history.

Still, other names are turning up today as well. "Reptoid" is a variation of the "Lizardman" label. The preference here will be to call these creatures by what appears to be a more descriptive phrase that correctly tells what these creatures represent. They are "Water Apes." They are our distant kin among the apes. More distant than many other intelligent primates, they took their own evolutionary path several million years ago, finding a very successful route through the geological ages. This book will explain why their success did not translate to a more dominant position in the world.

One day, water apes will be a familiar topic among humans once we can put the notions of "Lizardmen" behind us. Once we can discuss these creatures for what they are, we will have made a leap in our evolution. We will gain a greater understanding of the water apes and our extraordinary planet.

ENDNOTES

1 Gee, "Flores, God and Cryptozoology," *Nature* (2004).
2 "Short Sight Sizzler," *The Daily Colonist*, 13 June 1967.
3 "$25,000 Offered for Mermaid," *The Daily Colonist*, 15 June 1967.
4 "Mermaid Visits New Spot," *The Daily Colonist*, 23 June 1967.
5 Hall, *Living Fossils* (1999).
6 Hall, "Why Nothing Gets Solved in One Lifetime," *Wonders* 8(2).

Chapter One
The Lure of the Mermaid

THE INCONGRUITY BETWEEN THE FABLED BEAUTY of the mermaid and her fishy tail has created an enduring and much-utilized symbol. The fairy tale mermaid is universally known. Everyone thinks they would know one if they saw one. This widespread knowledge has established these creatures as one of the most familiar cryptids. However, these "unbelievable" features have made mermaids the very essence of incredibility to many people.

Philip Henry Gosse, an English naturalist, anticipated in 1862 the many mermaid speculations that would come in the 150 years to follow. Gosse was famous for creating the first public aquarium at the London Zoo in 1853, and also for coining the term "aquarium." Writing in *Romance of Natural History,* he made points that have appeared in later discussions about the antiquity of the creatures and how they have been explained away. The ancient world knew of merfolk. Gosse's earliest source was a Babylonian priest of the 3rd century BC:

> According to Berosus there came up from the Red Sea, on the shore contiguous to Babylonia, a brute creature named Oannes, which had the body of a fish, above whose front parts rose the head of a man; it had two human feet, which projected from each side of the tail; it had also a human voice and human language. This strange monster sojourned among the rude people during the day, taking no food, but retiring to the sea again at night; and continued for some time, teaching them the arts of civilised life.[1]

When we consider the antiquity of merbeings, we are inclined to allow that they might have accumulated wisdom. And perhaps one of these water dwellers reached out to humans with advice derived from their heritage and contact with humans in a prior age. Among the merfolk, knowledge passed from generation to generation will likely be as regional and variable as our own cultures.

If there is any truth to this remarkable account of the generosity of the water dwellers, human beings have been ungrateful in the extreme. A historical trend explains the distance between humans and the merfolk. A few thousand years ago, people withdrew from rural landscapes into city-states. They became more concerned with codifying knowledge and scheming with the politics of government and warfare. They were less and less in contact with Earth's creatures, like merbeings, satyrs, and other living fossils.. Those ancient folk began to doubt the reality of things they no longer saw for themselves. Into the Middle Ages, parts of Europe still knew the Trolls of the forest and the water-dwelling merfolk. But, as the world grew smaller, the interaction with such beings decreased. That diminishment continued for hundreds of years, to the point where the 20th century was satisfied that all those legendary creatures had been put to rest in the pursuit of the codifications of the sciences.

A few brave scholars have gathered accounts of merfolk. They turned up in the net cast by the venerable Charles Montgomery Skinner, an American fiction writer and journalist. His books of collected folklore are bursting with unusual animal reports that came to his attention during the 1800s, especially his twenty-five years as an editor at the *Brooklyn Daily Eagle*. Many contemporary accounts of lake monsters and the like would have been lost had they not come to his attention. He included them in his various books on "myths and legends" because they were not permitted to be seen as anything else in his day. In 1900 he ably summarized the divergence of the mermaid theme into humbugs, folktales, and arts worldwide in a brief but eloquent summation. It is too long to quote here, so it has been included as an appendix.

Skinner was the first in a series of serious attempts to collect mermaid accounts. U.S. Navy Lieutenant Fletcher S. Bassett assembled a colorful and far-reaching collection of merfolk stories that became a chapter in his *Legends and Superstitions of the Sea and of Sailors* (1885). Folklorists Gwen Benwell and Arthur Waugh gathered together the worldwide appearances of merfolk in art and folklore in *Sea Enchantress: The Mermaid and Her Kin* in 1961. Most recently, historian Vaughn Scribner has waded in on the subject. His *Merpeople: A Human History* believes the evolving representations of merbeings reveal insights into the ever-changing views of humans on religion, science, myth, and capitalism, lest we forget a particular major coffee chain with a mermaid logo.

In the 21st century, we are challenged to relearn what our distant ancestors once understood as fact. They knew firsthand that the satyrs were apes of a peculiar kind, true giants (often known as dangerous ogres) dwelled in the mountains, and trolls were hairy, muscular forest dwellers with secret knowledge that made them appear magical.

There was no place for these creatures in the world that human beings were building, first in city-states and then in the towns of Europe that grew around monasteries and markets. These beings did not own property, did not fight in armies, gave no loyalty to princes and kings, and did not live by the laws of any country. They were still encountered by simple fishermen, rustic farmers, and the occasional startled traveler. Instead of recognized zoological entities, they became staples in children's literature ranging from "The Little Mermaid" (1837) by Hans Christian Andersen to *The Sea Fairies* (1911) by L. Frank Baum.

From *The Little Mermaid* by Hans Christian Andersen, 1837.

As these creatures withdrew from open and frequent contact with human beings, they became even more mysterious in their nature and ways. But they lived in our folklore, the repository of all things not otherwise categorized. We have found the mermaids to be especially appealing. This allure probably goes back to the days beyond the ancient world. That earlier period might be called The Age Without Doubt. At such a time, our ancestors lived a simple life closer to the land and the waters. There was no doubt in their minds about the true nature of merbeings, satyrs, and others. Those creatures were all so numerous that all would have known them. The creatures might not have been as numerous as people or some of the game animals of the land. But as natural beings, they would have had their place in the scheme of things, and certain precautions could have been in force to allow them to co-exist with people. Those precautions are found in the very folklore that anthropologists have written off as quaint folk beliefs. They are relics of the Age Without Doubt, a time when people knew the inhabitants of the world better than we do today.

We can surmise that, thousands of years ago, the interaction between human beings and some of the merbeing cultures was sufficient to share our human conception of beauty with some of those cultures. In Greek folklore, this fascination with non-human primates is discussed under the headings of Nymphs and Nereids. The amount of material from Greece alone is extensive, both from the past and from stories told as recent events.

> The familiarity of the peasants with the Nereids is more intimate than can be easily imagined by those who have merely travelled, it may be, through the country but have no knowledge of the people in their homes. The educated classes of course, and with them some of the less communicative of the peasants, will deny all belief in such beings and affect to deride as old wives' fables the many stories concerning them. But in truth the belief is one which even men of considerable culture fail sometimes to eradicate from their own breasts ... As for the peasants, let them deny or avow their belief, there is probably no nook or hamlet in all Greece where the womenfolk at least do not scrupulously take precautions against the thefts and the malice of the Nereids, while many a man may still be found ready to recount in all good faith stories of their beauty and passion and caprice.[2]

Naturally, this information is considered doubtful and should not be taken as factual proof of humans making contact or having sexual relations with such beings. The sheer volume of reports on the subject over the years can make truth and fiction difficult to analyze. However, more analysis could be done if more reports of modern incidents with merfolk and satyrs were collected. Unfortunately, there has not been a surge of volunteers willing to pursue the task just yet.

A parallel exists with the satyrs because they too have females regarded as beautiful by human standards. Ivan T. Sanderson, a British biologist and nature writer, proposes that satyrs became known as the *wudéwásá*, the wild man of European folklore and British heraldry. He asserts that the *wudéwásá* inspired medieval architecture's "green man" motif (carvings of men's faces peering out of green foliage). By any name, Sanderson believes they are either surviving Neandertals or at least recollections of them.[3] The similarities to their multi-named water cousins are self-evident.

I realize this parallel will not be received well by those who regard satyrs as fabulous as mermaids. But it will bear some importance to those who will consider the evidence for these creatures. Our ancestors were in contact with satyrs and the merfolk, and we have lost touch with all that distant prehistory. Its traces have come down to us as common folk wisdom. We will likely continue learning from the neglected value of that folk wisdom in the coming centuries as we build a record for the evolution of all things around us. That evolution will be found to include these wonderful beings told of in folk stories all these years.[4]

The ancestry of the merfolk is not known for sure. But there is a good possibility that the large family of water apemen—found worldwide—are descendants of the primate family *Oreopithecidae*. The great grandfather to them all would be *Oreopithecus*. It's possible that water apes are the modern results of evolutionary changes to *Oreopithecus*. This fossil ape has been a topic of dispute. In the 1950s, paleontologist and zoologist Johannes Hürzeler suggested that *Oreopithecus* was a fossil hominin, an extinct member of the human family tree.[5] Before that, since its original description in 1872, it has been alternately labeled hominid, pongid (great ape ancestor), or cercopithecoid (Old World monkey).[6] The debate remains unresolved, but a recent examination of fossil teeth and jaw suggests that *Oreopithecus* was "a specialized semiaquatic folivore who apparently waded bipedally into freshwater swamps to feed on aquatic plants. However, the extensive wear on the oreopithecine canines and incisors, along with their manual precision grips, may indicate that freshwater invertebrates were also included in their diets."[7] In other words, this early primate had already adapted to a swamp environment, living on plants and small animals they could capture.

This echoes the work of English marine biologist and marine ecosystems expert Alister Hardy, who argued that the traditional savannah-dwelling origin of primates is incorrect. Hardy's original 1960 "waterside hypotheses" theorized that *Homo sapiens* exhibit physical traits more suited for an aquatic lifestyle, such as subcutaneous fat (insulation), voluntary respiration (holding their breath), and hairless bodies (reduced resistance when swimming).[8] Hardy further suggested as man gradually explored deeper water, he learned he could rest with his feet on the bottom and his head above water, standing erect with the water supporting his weight. As his balance improved, he found he could also stand upright on dry land, which led to chasing prey by running.

Marc Verhaegen of the Study Center Anthropology in Belgium revisits Hardy's theory with the benefit of decades of additional research.[9] Verhaegen suggests the earliest hominids waded in swamps and flooded coastal forests in Africa. There, they developed the anatomical and physiological features needed for survival in a water-adjacent environment rich in nutrients.[10] The aquatic apes may actually have evolved a more advanced brain first. Their diet was exclusively shallow water and coastal foods rich in brain-specific nutrients such as docosahexaenoic acid (DHA).[11] The ultimate source of DHA is algae and plankton; the primary consumers of algae/plankton are fish and shellfish. An ape with a diet including foods rich in DHA, such as fish, shellfish, turtles, and seabird eggs, will develop a more advanced brain quickly.

There was another benefit to spending time in the coastal waters. Pachyosteosclerosis is a thickening of the skeleton. This extra bone increases the animals' density and weight, which helps counteract buoyancy, allowing early water apes to remain submerged longer to feed while expending less energy. Some early hominids, such as *Homo erectus*, have been shown to have such an increased density in bones. Verhaegen believes these heavier, thicker bones would make it easier for early man to hunt in coastal waters for food sources and would compensate for the lack of stability from bipedalism.[12]

The difference between the Verhaegen waterside "aquatic ape" theory and our "water ape" is that Verhaegen believes that during the Pleistocene cooling, all remaining primate ancestors returned to dryer land as bipedal omnivores and scavengers. As the fauna adapted to new climate conditions, the primates became hunters focused on coastal prey.

We think some stayed in the swamps and avoided the competition for limited resources on the land. It makes sense that some would literally test the waters. *Homo sapiens* followed the coasts and rivers, providing food and the route for the rapid spread of man in the late Pleistocene. So did the water apes, but from offshore.

We have a curiosity about the ancestry of the primates. This kind of knowledge is not passed down from generation to generation. It can only be pried out of rock and soil. In that process, we find ways to date some ancient finds and compare them with care to show changes in their make-up. We are divining the origins of all the living fossils through this process, even as we refuse to acknowledge their surviving offspring.

The time may be right for some key reasons. One is that human beings are developing a context for the evolution of primates that is a long and busy picture. It contains many lines of intelligent creatures. Their origins date back to the Miocene. The lines extend all the way into our times, for the survivors are all around us, and they are no longer easy to ignore. They are Neo-Giants, Little People (*Homo floresiensis*), Trolls (*Homo gardarensis*), Neandertals, and *Homo erectus*.

Another reason would be that within the context of fossil finds, there are bones that could indicate the existence of water ape ancestors. But, of course, this will likely be an unwelcome and scarcely mentioned topic. Few people will be willing to go out on a limb on this subject. The bones of *Oreopithecus* are an excellent ancestor for the origin of water-friendly primates.

Many will want a clear-cut one-to-one correlation between fossils and the presence of water apes and Lizardmen. Furthermore, they will wish for recent finds of such fossils before backing the notion that these beings survive. Such discoveries come about in one of two ways.

First, there is the chance to find something not expected. This recently occurred with the evidence for Little People. Anthropologists were excavating a cave on the island of Flores. They were looking for ancient remains of *Homo erectus*. What they found were the very recent remains of *Homo floresiensis*. The ramifications of this are still not fully appreciated. Of course, these bones of Little People are not an isolated case of local dwarfing. They represent the kind of Little People that have come to inhabit much of the planet. They succeeded in proliferating and surviving, much like the water apes. They were early successes curtailed in their present state by the later success of *Homo sapiens*.

Another way for evidence to surface is for people to look for the specific fossil basis for a suspected creature. Such a directed effort would include examining fossil collections already made for some overlooked pieces previously found. *Homo gardarensis* is an example. A large jaw and skull fragments were discovered in a burial site at Gardar, Greenland, in a 12th-century Norse settlement. The "Gardar man" was briefly considered a surviving Neandertal until Sir Arthur Keith, a noted anthropologist, arbitrarily decided the oversized jaw belonged to a Norseman who suffered from acromegaly.

There's a strong argument that these bones are mislabeled.[13] They are allowed to lie neglected until some professional scientist tries to classify them correctly. Unfortunately, this misidentification made a century ago has caused the find to be overlooked in the Panum Institute in Copenhagen.

Is the time right for us to, at last, accept the existence of these miraculous merbeings? We can look back at the record of continuous encounters with mermen and mermaids. We can see that reports have continued through the 20th century, despite the universal displeasure expressed whenever writers have dared to suggest that these things should be taken seriously.

So let us start at the beginning, with *Oreopithecus*.

ENDNOTES

1 Gosse, *Romance of Natural History* 2nd series (1861), 125.

2 Lawson, *Modern Greek Folklore and Ancient Greek Religion* (1910), 131.

3 Sanderson, *"Things"* (2006), Chapter 10.

4 Hall, "The Satyrs in Our Midst," *Wonders* 9(3), September 2005; "Satyrs and Centaurs," *Wonders* 10(3), September 2006; "Satyrs in Africa," *Wonders* 10(3), September 2006.

5 Hürzeler, *"Zur systematischen Stellung von Oreopithecus,"* *Verhandlungen der Naturforschenden Gesellschaft in Basel* 65(1), 1954.

6 Delson, "An Anthropoid Enigma," *Journal of Human Evolution* 15, November 1987.

7 Williams, "Cranio-dental evidence of a hominin-like hyper-masticatory apparatus in Oreopithecus bambolii," *Bioscience Hypotheses* (2008).

8 Hardy, "Was Man More Aquatic in the Past?" *New Scientist* 7 (174), 17 March 1960, 642–45.

9 Verhaegen, "The Aquatic Ape Theory: Evidence and a possible scenario," *Medical Hypotheses* 16 (1), January 1985, 17–32.

10 Verhaegen and Munro "Pachyosteosclerosis suggests archaic Homo frequently collected sessile littoral foods." *Homo—Journal of Comparative Human Biology* 62(4), August 2011, 237–47.

11 Broadhurst, C. Leigh, et al, "Littoral Man and Waterside Woman: The Crucial Role of Marine and Lacustrine Foods and Environmental Resources in the Origin, Migration and Dominance of Homo sapiens," in *Was Man More Aquatic in the Past?* 2018.

12 Verhaegen and Munro, "Pachyosteosclerosis."

13 Hall, *Living Fossils: The Survival of Homo Gardarensis, Neandertal Man, and Homo Erectus,* 1999.

Chapter Two
Oreopithecus

LOOKING BACK AT THE RECORD OF PRIMATES, Oreopithecus stands out as the primary choice among hominoids that could have developed into the merbeings that survive into the twenty-first century. They are certainly not the only primate that took to the water for survival. Neandertal and *Homo erectus* skulls show ear exostoses, or "surfer's ear," an abnormal bone growth within the ear canal, typically caused by extended time in cold water.[1]

However, *Oreopithecus* displays a foot like that of no other primate. Its big toe sticks out at about a 90° angle from the remaining toes, shorter and straighter than the toes of living apes. The foot provided a firm base for an

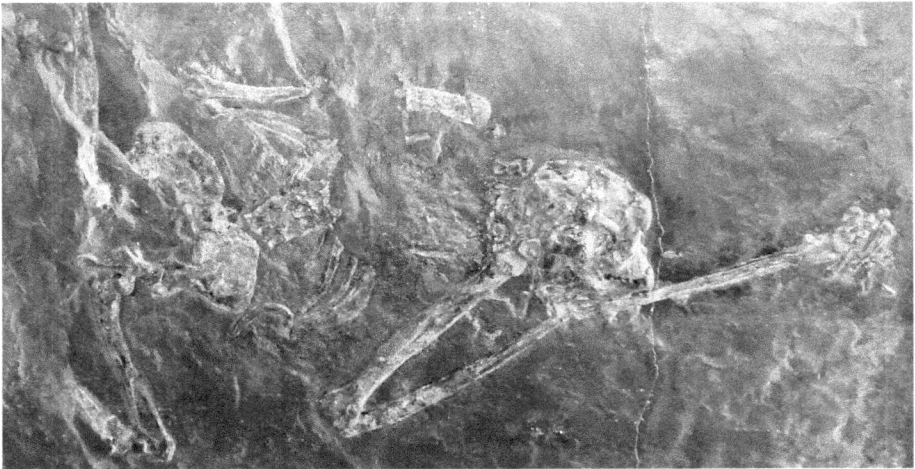

Fossil of *Oreopithecus bambolii*, known as the "enigmatic hominoid." Ghedoghedo/ *Wikimedia Commons.*

upright stance. However, its birdlike tripod design probably restricted the animal to a short, shuffling stride, optimal for the uneven marshy bottom of their home. The tripod-like foot of this primate may have developed into the four-toed foot of the North American freshwater apeman. Once at a ninety-degree angle, the big toe could have become the auxiliary toe indicated in footprints.

There are immediate issues with claiming water apes are mermaids. Mammals cannot breathe underwater, although an extended stay underwater might give that impression to a casual observer. For example, consider Cuvier's beaked whales, which can stay underwater for up to two hours.[2] The water apes could develop similar adaptations, such as more efficient respiratory system and increased oxygen storage in their myoglobin. Their ability to remain underwater would exceed a human, and they would appear to breathe underwater. The drawback is that increased myoglobin can be toxic to the kidneys. After severe trauma or muscle damage, myoglobin is released into the bloodstream, increasing the risk of kidney failure. This may explain why so many reports of merbeings are specifically about avoiding contact with humans with our historic "attack first, investigate later" mentality. It would also explain why so many stories of captured mermaids result in death in a short time.

We can't know for sure how these creatures adapted. However, it is short-sighted to assume that they cannot have evolved adaptations or developed their own technology to survive.

In old illustrations, some varieties of mermen and mermaids look peculiar, with long or double tails. Sometimes they are shown as having four limbs and a long tail. If such depictions are more than artistic license, the features could be simply cultural items. Consider them as costumes. They might dress up with different variations of their supplementary skins. Some of them might even be impractical costumes. Human beings certainly have come up with reasons to dress up and appear in all manner of impractical outfits. Our water cousins might do so as well.

We need to go beyond a view of these creatures as primitive beings that have always been primitive. Their presence as intelligent primates appears to go back tenfold before the appearance of *Homo sapiens*. They are not primitive creatures. They seem to be more advanced primates specialized to aquatic norms rather than the land norms we know.

Some of the reported features might be artificial. They are said to have yellow hair, green hair, and black hair. The yellow hair could be dyed. They might even sport wigs at times.

The pointed parts of the head observed in North America are sometimes compared to a rooster's comb. Still, they might have a more practical purpose. Maneuverability in the water is likely to be a key to the success of water apes, and the combs might assist that activity. This feature may have evolved as an extension of the sagittal crest, the ridges atop the skull, reported in some of the fossils of *Oreopithecus*.

We should not overlook the apparent origins of these creatures. They are likely to have the capacity to do many things that are denied to them in the current circumstances. Their lives are anything but static. Given the power of geological upheavals and the changes made to the planet in the last two hundred years, they have been put under pressure to adjust to many changes.

Mermaids have turned up along American coasts. Inland waters seem to be the home of freshwater apes of a different description that are definitely not described as a merbeing. They are tall, eight to nine feet in height. They are formidable creatures capable of great intelligence. They are torpedo-shaped and somewhat bow-legged out of the water, with brownish-black skin. They can even be mistaken for a tree or a log when standing or lying down. They exhibit pink or yellow eyeshine. For lack of a better name, we refer to these specific water apes as "lizard people."

They have four-toed tracks that show three primary toes and a trace of a fourth toe, which might even serve as a poison spur similar to that found on the platypus. If it does serve as a poison spur, then the slow loris may not be the only venomous primate. The loris has a toxin-producing brachial gland under its arm. The animal licks the gland because mixing the toxin with saliva creates a venomous bite.

Mermaids have been seen in freshwater wetlands from Maryland to Florida. Accounts come from the heartland from the Gulf Coast to the north into western Kentucky. Perhaps they are still around the wetlands in Great Lakes states such as Michigan and Wisconsin. They appear to inhabit the Pacific Northwest maritime coastline.

Widely scattered track reports suggest that merbeings are probably holding out where wetlands are still extensive enough to provide food and conceal from human interference. For the present, their very unexpected appearance will mean that if a human gets a glimpse of a lizardman, they will fear him and will not know what to make of the encounter.

People come up with whatever curious reasoning will allow them to minimize or dismiss the presence of such creatures. They think they are aliens, foreign frogmen, or survivors of freak accidents, or simply cannot be real.

Observations of individuals do not infer the surviving population. Mates and offspring would be kept out of contact with the dangerous keepers of firearms and diseases that might harm the apes. So we will not see them often and cannot say anything about them. Due to their habitat, their food sources would include fish, turtles, vegetation, mussels, birds, and eggs, just as *Oreopithecus* consumed millions of years ago. Furthermore, information suggests that merbeings in North America have a relationship with another creature: the alligator.

One of the continuing curiosities about alligators is that they continue to turn up in many places outside their range in the United States. However, the reports are such an insignificant occurrence that they fall below the radar of naturalists and the levels of government concerned with wildlife. There is the possibility that water apemen, in pursuit of their natural need for food in North America, engage in their own form of alligator farming.

Alligators typically occur as far west as Texas. They occur through the Gulf States and up the East coast into Virginia.

One popular notion almost invariably is voiced when alligators are mentioned outside this range. We are told that someone has released an overgrown pet into the natural environment. But can those instances explain all the alligators that are said to exist in the wild? For example, there is a swamp in Michigan where people insist that alligators can be found and are even nesting.

Living in the proximity of alligators could have advantages for water apemen. Three primary conditions of this situation come to mind. First, they could exploit the feeding habits of alligators. Alligators kill their prey and then store it in a riverbank to age before they return to devour it. The merbeings could regularly rob the food storage lockers of alligators for their own food. Second, in lean times, the alligators themselves could become food. Third, alligator hides might be useful in ways that only the merfolk know. After all, alligator skin has been a popular leather for many years, partially due to its sturdiness.

Perhaps the water apes have deliberately transported alligators to northern swamps to seed the wetlands near where the merfolk live.

A curious discovery by Archibald Rutledge, the first South Carolina poet laureate, shows how the relationship between merbeings and alligators could be less than peaceful. Among Rutledge's reminiscences of observing wildlife in the American South was one particular experience in South

Carolina. The incident occurred because he had permission to wander at will the 40,000 acres of the Santee Gun Club, a wild preserve.

On one of his excursions, he came upon what he said was "an enormous bull alligator, a regular primordial dragon." It was located in a pine glade more than a mile from reedy wetlands known as Blake's Marsh. It was dead when he saw it. His description of it was extraordinary. Rutledge wrote:

> From the tip of his burly snout to the end of his tail he was a perfect wreck. He was dead. His broad malignant head, his burly body, his massive tail—all were literally cut to ribbons. There were great gaping holes through his ponderous bulk.[3]

Rutledge thought the damage was inflicted by "some enemy, far his superior." Rainfall had removed any tracks of the combatants. However, signs of the struggle remained: vegetation was broken and matted. The bark on a nearby yellow pine had been torn away in battle. Rutledge went on to speculate on the victor's identity. He considered another alligator, a buck deer, and one or more razorback boars, but each possibility he found wanting. He could not connect the wounds with any wild creatures he knew to inhabit those woods.

Of course, we will never know what creature was the victor in that titanic struggle. The victor over the bull alligator might have been a male water ape, for the mermen of legend have been reported in many Southern swamps during the past two centuries. Perhaps this alligator was being kept as a food source or a source of leather but put up more of a fight than expected. Or maybe a merbeing was attempting to raid the alligator's food store and was ambushed. The brutality of the scene was so out of the ordinary that Rutledge concluded, "I only know that his assailant must have been terrible both in his strength and in his wrath. Only triumphant hate could produce such a malignant orgy of mutilation."[4]

While this particular attack may have been brutal, it would be unfair to make any generalizations. The abilities of the water apemen should not be judged and limited based solely upon the individuals reported, just as human actions and abilities can't be gauged upon the actions of just any one individual.

None of the higher primates get a free pass from the geological upheavals that occur every 30,000 years. We have the story of Oannes coming out of the sea to give knowledge to humans in the ancient world. This makes sense to us when we consider that the water apes may have accumulated knowledge over a long period of prehistory. Due to the periodicity of disasters, they would also have difficulty maintaining their own cultures.

Oannes as represented at the palace of Khorasabad, redrawn from McClintock, *Cyclopaedia of Biblical, Theological, and Ecclesiastical Literature*, 1868.

The true story of the mermen and mermaids will be more interesting if we view them in their proper context. They are not the bizarre anomaly they appear to be at first sight. They are a part of nature, much as human beings are a part of nature. The water creatures have made their way up an evolutionary ladder, just as humans have, from a more primitive type of primate that lived millions of years ago. And as discussed below, mermen have been vulnerable to the natural forces on this planet. These forces have hampered their progress and limited their success to the state we find them in today.

The prehistory of primates is filling in fast. The search for primate fossils has been ongoing for the last century. We have a patchy view of the previous twenty million years. Still, it is good enough to see how we and the other mysterious primates (characterized as "living fossils") fit into our current record.

Each of these prehistoric primates has living descendants surviving to the present. Each has a fascinating story in its own right, and the water apes are one among them. Although the mermen have their own line of descent from more primitive ancestors, they cannot and should not be viewed entirely apart from their fellow primate survivors. For this reason, it is required that we refer to the other primate survivors by name when discussing merbeings. This chapter will outline the larger picture of how merbeings came into being. Our knowledge about the evolution of merbeings is continually improving with each new fossil discovery. It will be subject to correction as we expand our understanding of our origins. We are constantly increasing the collections of primate fossils and the historical information about living fossils.

Naturally, we humans have been inclined to look for our own ancestry among the fossils dug out of prehistoric layers of rock. A time of significant importance to the story of modern primates was a period some sixteen million years ago in the Miocene Era when monkeys and apes were flourishing. It appears that many of the extraordinary survivors in the primate sweepstakes will one day be traced back to the fossil apes and monkeys of that period. Some survivors look like they will have evolved directly from one of those creatures. Humans, it seems, are descended from one of those creatures but share their ancestry with the other hominids among the living fossils. The best candidate for our ancestor is a fossil known as *Pierolapithecus catalaunicus*, which dates back thirteen million years. This ape later gave rise to four lines of descent, with humans being one of the four.[5]

Evidence currently suggests that the earliest upright-walking ancestor to humans dates around four million years ago. And the remains of the earliest humans fully formed as *Homo sapiens* date back to just under two hundred thousand years ago. But, again, these views are subject to change as more prehistoric fossil discoveries fill in the blanks. The physical forms and capabilities of these ancestors were changing drastically over those thirteen million years. We expect merbeings would be subject to similar changes over the millions of years that fill the time from the known fossil ancestor proposed below to the present day.

Pierolapithecus has been determined to be the ape that would have evolved into modern man and the three other close relatives, including *Homo gardarensis*, *Homo neandertalensis*, and *Homo erectus*.[6] This divergence of hominid lines has occurred over the last few million years. All three have living offspring still seen and studied in the flesh.

The bones of *Homo gardarensis* are evidence of a twelfth-century offspring of the line of descent from *Homo heidelbergensis*.

The Neandertals have living representatives also. Their fossils are definitely on record in Europe and Asia, but any modern survivors in Eurasia are uncertain. However, a relatively recent population made their way into the New World. They are still found in Alaska and parts of Canada.

Homo erectus is still found across parts of Central Asia, Pakistan, and China, based on numerous sightings and captures in those parts of the world.

Relations more distant from humans are also modern survivors. Another line of descent has produced the primates known as the Neo-Giants of Asia and America (more familiarly known as Bigfoot and Sasquatch). The Neo-Giants have the characteristics of *Paranthropus*. Further, there are the Little People (*Homo floresiensis*). The Little People have been around so long that they are found in many places worldwide. They are not confined to Southeast Asia, as hypothesized by those who first discovered their skeletal remains. As more is ascertained from their bones, we will learn they are likely descended from Australopithecines. Some of their attributes already suggest this.

The True Giants are still another line of descent. Their ancestor can only be the largest primate known to have existed, *Gigantopithecus*. This identification has been hindered by paleo-anthropologists claiming this fossil was a semi-erect ape. The record of True Giants in Malaysia and elsewhere has established their survival into modern times. It is possible that new finds of bones and fossils will settle this issue in favor of the erect True Giant that so many people have seen in the flesh in recent decades.

Another ape of some importance is *Dryopithecus*. It is an early fossil ape and a large one. This ape appears to have two modern forms: the Yeti and the Satyr. Today, their preference for wetland habitats suggests they may have favored those areas in the early competition with other land primates. In that case, they would have been competitors with the earlier forms of merbeings.

In recent years, some modern descendants have come into prominence in the news around the globe. This is true even though our modern primate cousins persist only in small numbers today. The discovery of the fossils of *Dryopithecus* and the reports of its kind turning up in many other places combine to tell us that this animal was dispersed far and wide. Dryopithecine apes have come into the news in Asia as the famous Yeti, whose tracks were photographed by the British mountaineer Eric Shipton and some others.

The rare dwellers of the Mississippi River bottomlands are also coming into prominence. Known as the North American Apes (or NAPEs), cryptozoologist Loren Coleman has called attention to them for more than fifty years. In the news, African apes, dubbed the "Lion-Killers," have been seen, their tracks cast, and a photograph shows a dead specimen.[7] They have also been highlighted by Coleman in *The Field Guide to Bigfoot*. All these primate survivors are part of the story of how *Oreopithecus* could have evolved into the water apes that still survive in many parts of the globe.

The numerous and diverse evolution of these primates created pressure in the environment. The combined stress of competition on land and swampy habitats are the possible forces that drove one particular species of ape to enter the water world around ten million years ago.

Oreopithecus is the most likely candidate for the ancestor of the water ape. This species of ape is known from fossils dating back to around nine million years ago. Identified from fossils first found in the 1870s, *Oreopithecus* was primarily a swamp dweller. Its remains have turned up in Italian coal mines. However, it is worth noting that one of the most complete fossils that turned up in 1958 was found in what appears to be a swimming position.[8]

Some fossils have been found in circumstances suggesting an island environment. It has been observed that the build of *Oreopithecus* seems to have been unsuitable for swimming. Still, no one suggests that Oreopithecus was born to be a "sea ape." Instead, it appears that in competition with the apes of its day, especially *Dryopithecus*, the ancient fossil form of *Oreopithecus* might have found an advantage in taking to the water. That early form of ape could reach safety on islands. Later, they would have found it beneficial to remain in the water for long periods.

Over millions of years, this ape might have adapted to freshwater and marine worlds to become the mermen and mermaids that have seen in recent centuries. Aquatic environments would routinely dispose of the remains of such creatures, so fossil finds of the descendants of *Oreopithecus* would be rare.

In the middle of the Miocene, we see increasing competition for space and resources on the land among the apes. However, the course of that competition will be better defined by fossil records yet to be found. We can already see there were a lot of primates on the land. To avoid direct competition with apes like *Paranthropus*, the Little People, True Giants, and *Pierolapithecus*, it appears that some apes sought refuge in the less desirable swamp terrain.

Oreopithecus is thought to have been a swamp ape. We know from modern accounts that *Dryopithecus* favors such areas in North America but is not a swamp dweller globally. So it appears plausible that these two creatures might have been in direct competition long ago. Facing the brutish physical strength of the Dryopithecines, the ancestors of today's water apes might have been pressured into taking to the water, first to reach the safety of islands and then adapt to an aquatic environment where there was more food.

The forms of *Oreopithecus* became physically adapted to a water world in body shape and limbs. They adapted or devised technological means to remain in and under the water for long periods. The diving bradycardia reflex, commonly called the diving response or mammalian dive reflex (MDR), is likely to have played a part in the evolution of water apes. The MDR is recognized even in human beings. It is regarded as a survival from the fetal life when the developing human has to survive in a fluid environment.[9] The MDR reacts to cold water causing a slower heartbeat and reduced blood flow.

Building upon such natural capacities, *Oreopithecus* could have evolved over millions of years to stay underwater for extended periods. Mermaids using skins, such as seals or fish, can be seen as tools to assist their success. They are no more surprising than the development by humans of swim fins, goggles, and wetsuits.

When the marine merfolk come out on land, they are reported to take off their supplementary skins and get about without them. By accounts that have become legendary, we are given the impression that such a skin or wetsuit can be considered vital for survival in the water of those particular individuals. They might need to regulate their body temperatures or increase mobility to avoid the hazards of their environment.

Likely, such a suit is not a necessity for all the merfolk. Here again, this large family of creatures have a variety of adaptations. Some may not have such restrictions because their bodies are fully adapted to their needs without a suit.

This numerous and successful family of primates may one day be recognized as diverse forms of similarly specialized primates, all in the family of *Oreopithecidae*.

What seems possible is that over nine million years, *Oreopithecus* was finding its way both in the world of inland bodies of water and in marine environments. As it did so, it branched into diverse forms of water-dwelling primates. They began as an ape, much like our ancestors from long ago. It should be noted that the beings that evolved were suited to water but could still come out of the water. Their evolutionary path would have favored certain traits that made them formidable survivors. They had to develop more efficient lungs to spend increasing lengths of time under the

Derceto, from *Oedipus Aegyptiacus* by Athanasius Kircher, 1652.

water without resurfacing for oxygen. Furthermore, swimming speed and maneuverability would have been vital to avoiding predators. Also, they could have developed weapons suitable for fending off predators and obtaining prey for food.

When we consider water apes as another highly evolved primate on this planet, we realize that they are likely to share many things with us in their nature. They are arguably the most successful creature in the water world. What separates them from us is their highly specialized appearance. Outwardly they will at first appear bizarre to humans. Once we understand who they are and how they have survived, we will likely find them entirely capable of thinking like us, manipulating their environment as we do, and yearning for the same basic needs of life.

The water apes have been hampered in their progress by the inherent instability of Earth's hydrosphere. Our planet is subject to periodic global changes. We have come to hypothesize this geological record as True Polar Wander. This hypothesis is still under constant examination, evaluation, and argument. However, it has not risen to the level of a theory. Still, its value in explaining the frequent changes in past environments is impressive.

According to True Polar Wander, the thin crust of the Earth shifts periodically upon the great liquid mass of the inner layers of the planet. Each shift involves the continents and oceans moving as one to a different point on the Earth. The magnetic poles remain in the same place but are now positioned over other geographic points. The extent of a shift is exemplified in the last theorized shift. Before that shift, Hudson Bay was at the geographic North Pole. Some twelve thousand years ago, the crust of the Earth shifted, moving that land area from a polar position to its present more southerly location. A similar shift was occurring at the South Pole. The effect was to free the underlying Hudson Bay area from its ice cap and to allow the subsequent post-Pleistocene invasion of that area by the bay and bands of vegetation and wildlife for that new latitude for the bay area. The periodicity of such shifts is still debated. The best estimate is that such dramatic shifts occur about every thirty thousand years.

Whatever the pace of this crustal movement, there appears by the evidence of paleogeography to be a sudden final shift to the new positions for the continents and all that surrounds them. Naturally, our discussions focus on the continents because they are our significant reference points. However, the Earth's entire surface contents, including the continents and the water environments we know as the hydrosphere, will all be in a state of continuous motion. They all end up in new latitudes. Our measuring system is tied to theoretical polar locations upon a continuously spinning Earth. Those polar points are stable. They do not shift; only the land areas and waters move in relation to the polar points. Then the life forms on those continents and in those bodies of water all have to adjust quickly to the new regimes of climate and vegetation or die.

True Polar Wander's impact on life is one of the most influential factors in creating the world we see today, with all its variety of living things. I would suggest it is also one of the significant factors in creating a world that contains so many cryptids. These creatures have survived by finding a niche in each new ecosystem created by these periodic disasters. Those cryptids are unknown animals that are rare and reclusive, just like merbeings.

It should be mentioned that there is a currently popular notion that meteoritic impacts have had devastating influences on life on Earth. Many others have remained skeptical of these proposals. It has always seemed to me that it is too simple to take a point in time in the past and then send a lot of academics scurrying to find evidence of an impact at a specific yet unknown point in time. Evidence of multiple impacts seems to be common in our long prehistory. It seems easy to find some record in the vicinity of any point in the distant past if one sets people looking at one time.

As part of the shifting of the crust, there will be geophysical effects on the planet. Periods of extraordinary volcanism have the potential for materials to burst forth from within the crust, create lava beds and the like, and for significant movements of continents and parts of continents. Mountain-building episodes are likely to date from these events. These things will be cataloged and put into a sensible framework as the records for True Polar Wander are studied in the future. These physical upheavals will have devastating consequences for all living things. So it seems that the effects of meteoritic impacts now being fashionably discussed will one day be replaced by a better idea. Those are the powerful forces unleashed in episodes of True Polar Wander.

Now we come to True Polar Wander's impact upon creatures like the water apes. Water-dwelling primates would have had a hard time creating a lasting civilization that relied upon their improvements to their physical environment. Periodic catastrophes would involve devastating changes to the water world. The advances within the intellects of the water apes and their physical adaptations would not be affected. Their habitat and any artifacts they might have developed would be destroyed or swept away amid extreme environmental upheaval.

Also, the impact on the water ape population may have been decimated at each occurrence. They would have to recover and rebuild over a long period of time. They would have to find new, more hospitable surroundings. Then they would increase their population again, only to have a similar upheaval occur after a comparatively short time. Thirty thousand years is a short interval in the geological records.

To be sure, the impact of crustal shifts will have been significant for other primates. We are slowly discerning the effects of these shifts upon the advance of human beings over long periods. And the effects upon others will be found if we look for them in future fossil records. But the water apes would have been especially vulnerable because they live in an aquatic environment. They swim in the very substance that would have been moved into convulsions by the power of crustal re-orientation.

Upheavals would also be on the land surface but to a much lesser degree. The traditions of some Western Australian aborigines seem to tell of just such conditions in the wake of a crustal shift. Such a tradition tells how for a day and a night, the Earth shook, and the next day the people found the landscape altered as described this way:

> It seemed as if the earth had turned round. The hills had moved. Where there used to be hills now there were plains and where there used to be plains there were now hills. Great rocks had risen along the coast. Powerful winds sprang up. The air was full of smoke and dust. This continued so long that many people died.
>
> Finally, fresh air began to blow. There was thunder and torrents of water rushed across the low-lying lands. Now the people who had survived ran to the higher ground.
>
> The strangest thing was that after the great shaking it seemed as if the earth had turned and the Sun was rising in a different place. Just as the hills had moved, so now the Sun that used to come up in the north and set in the south was rising in the east and setting in the west.
>
> After this time the people divided into bands or tribes and moved into different areas, and they also developed different languages. The medicine men were respected and could travel between the different tribal areas.[10]

Of course, it is not the heavens that would be changing at that time. Instead, the continent would have changed its orientation so that the people's reference points would have altered in relation to the heavens.

The impact was significant on the surface. It is probably mild compared to the disruption suffered by water apes worldwide by turbulence, underwater avalanches, changing water levels, and coastal changes during the same period. In addition, the impact on coastlines would affect one of the prime habitats for water apes.

The success of the ultimate swamp ape can be seen to have been widespread in their adaptations to freshwater and marine environments worldwide. The traditions of them among human populations in recent history attest to this. Their problem in advancing their lot as a cooperative effort among their kind of primates would have been a geologically constant interruption of their wellbeing by devastating natural forces. In this fashion, they are distinguished from the other higher primates. The other creatures were competing with each other over space and resources on the land. The water apes held sway over the hydrosphere. They were unconcerned with issues historically affecting higher primates such as humans, the Trolls (Marked Hominids), the Neandertals, the Little People, the Neo-Giants, the True Giants, and others.

Over the last few thousand years, the recent history of human beings has created concern for water apes. We humans are now altering the landscape to suit ourselves. We drain swamps, alter wetlands, and even destroy entire ecosystems while thinking we can create new ones elsewhere to compensate for the ones removed. We drown some swamps to create new lakes; we drain lakes; we change the course of rivers and erect dams to suit our needs. We are more interested in coastal regions, using them as we see fit for occupation and industry. Such areas were isolated and unused in the distant past, except for specific desirable ports. So humans have become a concern for the water apes today, much more so than we would have been during past epochs.

Is the world ready to face the existence of merbeings? There will be the usual resistance to giving up on the comfort of ignorance. Expect the usual complaints that we are not critical enough. "Everyone makes up stories," they will say. "Isn't it peculiar that so many teenagers see these things?" they will ask. No, young people have usually been the ones out at night, going camping and fishing, and so running into such things as these.

But how could there be so many different-looking beings exist undetected? This is primarily because people have declined to apply advanced technology to seriously look for them. Ergo, they remain out of sight until we try to find them.

ENDNOTES

1 Kennedy, "The relationship between auditory exostoses and cold water: a latitudinal analysis." *American Journal of Biological Anthropology* 71 (4), (December 1986), 401–415.

2 Quick et al, "Extreme diving in mammals: first estimates of behavioural aerobic dive limits in Cuvier's beaked whales," *Journal of Experimental Biology* 223 (18), September 2020.

3 Rutledge, "Mysteries of nature: peculiar occurrences that defy explanation," *Nature Magazine* 27(5), May 1936.

4 Rutledge, 265.

5 Culotta, "New Spanish fossil sheds new light on the oldest great apes," *Science* 306(5700), 19 November 2004.

6 Moyà-Solà, "*Pierolapithecus catalaunicus.*" *Science* 306(5700), 19 November 2004.

7 Young, "The Beast With No Name" *New Scientist* 184(2468), 09 October 2004.

8 Fleagle, *Primate Adaptation and Evolution* (1999), 471–2, 474.

9 Goksör, "Bradycardic response during submersion in infant swimming." *Acta Paediatrica* 91(3), March 2002.

10 Wolfe, "The Great Shaking" *Flying Saucer Review* 40(2), Summer 1995.

Chapter Three
North American Native Accounts

THE LEGENDARY ACCOUNTS OF WATER APES from the Indigenous Peoples of North America are primarily from the 20th century. The water apes' place among the Native American cultures suggests that these creatures have existed a long time. The incidental accounts of merbeings have a substantial, long-term history, even though they may not be as numerous.

The Penobscot people's *alambaguenosisak* (under-still-water-little-folk), or *lumpeguin* for the Passamaquoddy, are considered a water spirit more akin to sprites than mermaids. Still, the *alambaguenosisak* stories include so many different-sized creatures that they are believed to be shape-shifters. They exhibit typical mermaid behaviors that appear across the globe with descriptions that cannot be written off as simply cross-cultural exchanges with the local British colonists.

The *alambaguenosisak* are considered saltwater creatures who are encountered in rivers and lakes. In Chapter Seven, we discuss why this may be preventative medicine. These merbeings have a tradition that if they are trapped on the land, they are under the control of anyone who steals their garments. As we examine our oreopithecine distant cousins, we see this recurring clothing-theft theme worldwide. This makes sense if you assume the water apes have developed protective wardrobe against the cold water, which also helps their speed and maneuverability.

One of the oldest records originated in the *Lewiston Journal* in Maine. A correspondent for the paper interviewed Father O'Dowd, a priest who was the pastor of the Passamaquoddy at the Pleasant Point settlement. Several articles appeared in the Lewiston paper as Father O'Dowd attempted to familiarize the Maine community with the people who lived there first. The American Folklore Society reprinted O'Dowd's material in a regular feature in their journal called "Folk-Lore Scrap-Book."

According to one scrapbook entry on Passamaquoddy superstitions, O'Dowd discusses the "Lam-peg-win-wuk." This is Father O'Dowd's attempt at a phonetic spelling of *alambaguenosisak*. O'Dowd explains them as "sprites who live under the water and sometimes dance in the waves. It is probable that these are really the phosphorescent gleams made by animalculae in the sea."[1] But O'Dowd may have been premature in his dismissal. As we will see in Chapter Seven, there are merbeings and lizardmen associated with bioluminescent algae across Asia.

O'Dowd also brings up the larger and more vicious mermaid of Passamaquoddy lore, the *apodumken* or *apotamkin*.

> Aboo-dom-k'n is an evil sprite that is believed to live in the water, to cast evil spells upon Indians who may stroll along the shore, or even to seize and devour children who may be playing in the water. Aboo-dom-k'n is supposed to have a fish's body and tail, with a woman's head and hair, and corresponds to our idea of a mermaid, if we have any.[2]

Also known as *nodumkanwet* in Penobscot, the *apotamkin* is occasionally called a giant serpent, a theme we see globally and particularly in Africa. However, in North America, the *apotamkin* became a "nursery bogey," used to frighten children into avoiding thin ice in the winter and unguarded beaches in the summer.[3] And while there is no shortage of such cautionary bogies in Native tales, very few are half-fish.

The *apotamkin* may have more in common with the African branches of the oreopithecine family tree. Usually, the *apotamkin* dwells off the coast of Passamaquoddy Bay. This gives the *apotamkin* access to the Bay of Fundy, between Maine and New Brunswick.[4] Navigating the Bay of Fundy is still difficult and dangerous, with "numerous off-lying dangers fringing the approaches, by rapid and uncertain tides, as well as by the frequent occurrence of dense fogs."[5] In the age of sail, before radar, sonar, and accurate charts, it was a high-risk passage. In other words, when the *apotamkin* wasn't stalking people along the beach or dragging unsuspecting young children into the water to devour, there was always a potential shipwreck with sailors to drown.

The Inuit of the Canadian Arctic tell similar stories of the *qallupilluit* (or *qalupalik*). Franz Boas, a German-American anthropologist known as the "Father of American Anthropology," repeated the account of the feared "kalopaling" in his studies on Baffin Island, the largest island in Canada.

> His body is like that of a human being and he wears clothing made of eider ducks' skins. Therefore he is sometimes called Mitiling (with eider ducks). As these birds have a black back and a white belly, his gown looked speckled all over. His jacket has an enormous hood, which is an object of fear to the Inuit. If a kayak capsizes and the boatman is drowned Kalopaling puts him into this hood.[6]

The hood that Boas refers to is an amautik, the parka worn by Inuit women with a pouch for carrying babies on their backs. The *qallupilluit* also serves as a "nursery bogey." Children are warned about going too close to the shore lest the *qallupilluit* carry them away in their amautik.

The *qallupilluit* has long hair and fins on its head and back. Instead of a tail, it has oversized feet that look like floats made of seal bladders attached to Inuit harpoon lines. This suggests some buoyancy control is needed to compensate for the insulated wardrobe in the arctic Atlantic Ocean.

Its hands are webbed and clawed. It reportedly has a distinctive smell reminiscent of sulfur. This is a common observation about cryptid hominids, which we'll discuss in Chapter Nine.

The *apodumken* and *qallupilluit* sound similar to the opportunistic scavenger mermaids like the *mambu-mutu* or *chitapo* of Africa's Copperbelt region (see Chapter Eight), who consider a human corpse just another protein source. This may explain the *apodumken* was frequently sighted in the Penobscot River near Bangor during the 1849 cholera epidemic [7]

Not all merbeings in the Northeast have antagonistic relationships with Native Americans. American anthropologists Ruth and Wilson D. Wallis did fieldwork writing down the experiences of the Mi'kmaq of Eastern Canada. Their collected stories of "mythic peoples" include the "Halfway People," also known as the *Sabawaelnu*. According to one interviewee:

> There are many Indians everywhere, all over the world, even in the bottom of the bay. In the halfway place the lower part of the body is like a fish, whereas the upper part is human. Sabawaelnu, "water dwelling people," is the name of these people. Sometimes they come up to the surface and ask for tobacco or a knife. If they are treated

well, they do no harm; but if they are treated badly, they will raise a big storm, and do great injury. Before a storm, you can hear them singing; and when they cease, the storm breaks.... When the storm is over, they resume their singing. Not everyone hears them singing, and not everyone has seen them. A great many of the old people, but not many of the younger generation, have seen them. Some of these people are male, and some are female.[8]

The Mi'kmaq had specific details that described encounters and dates for those meetings that go back to around 1875. The above informant had once found a child of the Halfway People. He dropped the child, and it fled immediately into deep water.

The *Sabawaelnu* were known to sing to warn the local fishermen that a storm was coming. The Mi'kmaw who angered the Halfway People drowned, while those that stayed in their good graces had great success with their fishing. These merbeings did not control the weather as the Mi'kmaq claimed. The *Sabawaelnu,* as an aquatic creature, would notice the behavior change in the fish. When there is a storm approaching, the barometric pressure begins to fall. An unwritten rule among fishermen is that when a storm is about to arrive, the fish sense a drop in barometric pressure and increase their feeding.

The success in fishing (or survival) may have more to do with the timing and severity of the approaching storm and the awareness of a Mi'kmaw when to come ashore. The *Sabawaelnu* signaling the village of the storm's approach could demonstrate a working relationship between the two peoples.

Charles Montgomery Skinner, a journalist known for his published collections of myths and folklore, took note of merfolk accounts. His book *American Myths and Legends* has numerous examples of the stories told about them.

An Alaskan tribe tells that it crossed the sea under the lead of a man-fish, with green hair and beard, who charmed the whole company with his singing.

The Canadian Indians relate that a member of the Ottawa tribe, while lounging beside a stream, was confronted by an undoubted mermaid that arose through the water and begged him to help her to the land. Her long hair hung dripping over her shoulders, her blue eyes looked pleadingly into his. Would he not take her to his people? She was weary of being half a fish and wanted to be all human, but this might be only if she wed.[9]

The Ottawa took her home, adopted her, and arranged her marriage to a member of the Adirondack tribe. Her presence led to a war between the Ottawa and the Adirondacks, leading to the eventual extinction of the entire Adirondack tribe. She was last seen in the Mississippi near St. Anthony's Falls, having turned back into a mermaid.

In his *Aborigines of Minnesota*, American geologist N. H. Winchell noted that the term "merman" was one of a few nicknames for a totemic grouping among the Ojibwe.[10] He reproduced two drawings of merman clan totems initially drawn in the 1880s by Herman Haupt.[11]

Merman clan totems among the Ojibwe. From *The Aborigines of Minnesota* by N. H. Winchell, 1911.

Albert B. Reagen, an anthropologist who specialized in Native American history, heard from the Ojibwe (Chippewa) people at Nett Lake in northern Minnesota about merbeings centered on Picture Island.

> Picture Island is in Nett Lake near the Indian village of the same name not far from Orr, Minnesota. It is low and its northern and eastern surfaces are polished rocks dipping into the placid water, the polishing having been done by glacial action during the Ice Age. In these sections its rocks are covered with crudely-made pictographs of human beings, dance scenes and outlines of the animal gods worshipped by the men who made the pictures."[12]

Reagen credited "some of the very oldest medicine men" for this legend, where the first Chippewa Indians discovered creatures at Nett Lake:

[The Chippewa Indians] had been canoeing only a few minutes in that lake when they came in sight of Picture Island, and lo, it was swarming with a multitude of beings that were half sea lion and half fish. On their approach, these became panic stricken and, fleeing to the west side of the island, took to the water and swam with all speed across the lake and up the creek that leads southwestward. Reaching the head of the stream and still being pursued, they dived down into the earth; and now the water bubbles forth from the place where they disappeared, a site still held sacred by our people. On coming to the island, the canoe-men paddled around it. Then by the track of the muddied water they pursued the beasts across the lake and up the creek till they found where the earth had swallowed them up as though they had been caught in a net. Since then we have the lake "Netor-as-sab-a-co-na" (Nett Lake, that is, the lake with a net). Then when our people had returned from the pursuit they found these pictographs on the island. They are the writings of those half sea lion, half fish beings.[13]

Reagen had his own interpretation to give to this story. He thought it was related to the aftermath of a battle between the Ojibwe and the Sioux (Dakota). A battle was known to have occurred eastward at Elbow Falls in the Vermillion Lake region. He thought some defeated Sioux had been seen on Picture Island and gave chase. His conclusion: "The Chippewa pursued them to the head of the stream where it has its course in a number of bubbling springs. That the pursued had had the earth swallow them up and that they were half sea lion and half fish beings was, no doubt, invented to account for their precipitous flight and disappearance."[14] Reagan's dismissal of Native accounts illustrates how multiple records of water apes have been explained away over the years.

Other notice of merbeings had been taken in Ohio where philologist and ethnologist Albert Gatschet recorded the names given to merbeings among the Native Americans in Miami. Mermen were called *mänsanzhi* and mermaids were called *mänsanzhi kw*ä.[15]

Amateur anthropologist and historian David Bushnell visited with some of the few remaining Choctaw people living north of Lake Pontchartrain in Louisiana. He heard how they associated merbeings with localities north of the lake. Like Father O'Dowd in Maine and Wilson D. Wallis in

Canada, Bushnell found his informants told of "things" they had encountered. These things were considered natural inhabitants of the region the Native Americans had occupied for generations. They told him of creatures they knew to live in the swamps and waters of the area. Among them were the *Okwa Naholo*.

> The Okwa naholo ("White People of the Water") dwell in deep pools in rivers and bayous. There is said to be such a place in the Abita river; the pool is clear and cold and it is easy to see far down into the depths, but the surrounding water of the river is dark and muddy. Many of the Okwa naholo live in this pool, which is known to all the Choctaw.
>
> As their name signifies, the Okwa naholo resemble white people more than they do Choctaw; their skin is rather light in color, resembling the skin of a trout.
>
> When the Choctaw swim in the Abita near the pool, the Okwa naholo attempt to seize them and draw them down into the pool to their home, where they live and become Okwa naholo. After the third day their skin begins to change and soon resembles the skin of a trout. They learn to live, eat, and swim in the same way as fish."[16]

Bushnell also recorded the story of someone nearly pulled down by what were known as the "White People of the Water."

> Heleema (Louisa), one of the women living at Bayou Lacomb, claims that when a child, some forty years ago, she had an experience with the Okwa naholo. She related it with the greatest sincerity. One summer day, when she was seven or eight years of age, she was swimming in the Abita with many other Choctaw children. She was a short distance away from the others when suddenly she felt the Okwa naholo drawing her down. The water seemed to rise about her and she was struggling and endeavoring to free herself when some of her friends, realizing her danger and the cause of it, went to her assistance and, seizing her by the hair, drew her to the shore. Never again did the children go swimming where this incident occurred.[17]

In British Columbia, the one to be feared was *Tchimose.*

> Tchimose, one of the deities of the Haidah Indians of Queen Charlotte's Island, British Columbia, who resides in the sea, has a human face and two tails. He wears a hat— symbol of a powerful deity—and is to be feared for his destructive tendencies towards canoes and their occupants.
>
> Another tradition of the Indians of British Columbia— that of the Lillooet tribe—concerns a people called Sainnux, who live in underground houses, and are at home in the water and powerful in magic.[18]

In Alaska, an 11-year-old Aleutian child reported encounters with 'ítcanam maya'ú, the Creek Man. Creek Man is a creature embodying both human and animal physical characteristics. Rather than malicious, Creek Man is more of a practical joker.[19]

The Southern Miwok people have written about central California's Merced River, home of the *ho-hā'-pe*, river mermaids.[20] Once the river runs into Yosemite Valley, the Ahwahnechee people similarly encounter the *ho-hā'-pe.*[21] These river mermaids protect their river zealously, and stories abound of encounters that end in tragedy for the Native American interlopers. This may not be fair to the *ho-hā'-pe.* The Ahwahnechee themselves were in a battle against the United States government to control what would become Yosemite National Park, which may have colored their accounts.

The *ho-hā'-pe* prefer deep freshwater pools, and the Miwok differentiate them from the ocean-dwelling *Le'-wah ke'-lak.* The Merced River joins the San Joaquin River, emptying into the ocean three hundred miles west in the San Francisco Bay. There are few reports of mermaids in the bay; as with many other locations, man's ships have driven the water dwellers to less trafficked areas.

A basis for the water ape can be found in both Native American lore and lore from other cultures within the Americas. It has been written down by anthropologists and overlaps with topics in American cryptozoology. Such is the case for the presence of merbeings. The Native Americans say they have known them as a natural part of their environment. Native lives were entwined with the natural world, unlike the city-building new arrivals from Europe and elsewhere of recent centuries.

When people ask, "What do Indigenous peoples say about the water apes?" the answer must be pieced together from notes taken during the last one hundred years by anthropologists, journalists, and folklorists. It was not the custom for the natives to compose nature handbooks. They passed along their knowledge directly from one person to another by word of mouth. The reality of an encounter was not questioned when you or your friends saw such things in the ordinary course of time. If a story about the dangers in the woods could serve as a cautionary tale, that suited them just fine. If we want to add water apes to our illustrated books about the true wonders of nature, we have a lot of work to do.

ENDNOTES

1 "Superstitions of the Passamaquoddies." *Journal of American Folklore* 2(6), July-September 1889, 230.

2 Superstitions of the Passamaquoddies," 229.

3 Rose, *Giants, Monsters, and Dragons* (2000), 271.

4 Hawk, "Apotamkin" in American Myths, Legends, and Tall Tales (2016), 41–42.

5 Orr, *Sailing directions for Nova Scotia, Bay of Fundy, and South Shore of Gulf of St. Lawrence* (1891), 16.

6 Boas, *Central Eskimo* (1888), 620.

7 Packard, *Mythical Creatures of Maine* (2021), 87.

8 Wallis, *The Micmac Indians of Eastern Canada* (1955), 349.

9 Skinner, *American Myths and Legends* (1903) v.2, 332–33.

10 Winchell, N. H. *The Aborigines of Minnesota* (1911), plate VII (p. 605).

11 Haupt, [papers concerning North American Indians], Newberry Library, Ayer MS 366.

12 Reagan, "Picture Island," *The Southern Workman* 55(10), October 1926), 457.

13 Reagan 459–60.

14 Reagan 461.

15 Gatschet, "Water-Monsters of American Aborigines," *Journal of American Folklore* 12(47), October-December 1899.

16 Bushnell, *The Choctaw of Bayou Lacomb* (1909), 31.

17 Bushnell, 31fnB.

18 Benwell, *Sea Enchantress* (1965), 208, 209.

19 Krukoff, "Aboriginal Ghost, 'ítcanam maya'ú, Creek Man." *Journal of American Folklore* 60(235), January-March 1947.

20 Merriam, *The Dawn of the World* (1910), 228-30.

21 Wilson, *The Lore and the Lure of the Yosemite* (1922), 125–26.

Chapter Four
North America

WHEN THE FIRST EXPLORERS ARRIVED in North America, they found monsters. New species and new dangers disrupted the complacent superiority of the Europeans. This forced them to describe unknown animals. Some were ordinary plants and animals, filtered through their familiarities and beliefs. Others are less simple to explain away.

Mermaid in St. John's Harbour by Theodore de Bry (1528–1598), courtesy of the Centre for Newfoundland Studies.

In 1610, an English colonist, mariner, and writer named Richard Whitbourne was on the shore of St. John's Harbor in Newfoundland. He and a companion saw something swimming swiftly toward them. It appeared to be a cheerful woman with a well-proportioned face. But as it swam closer, they realized it had no hair, just neck-length blue streaks resembling hair. When it came within twenty-five feet, they began backing away from the water. The merbeing saw them putting some distance between themselves and the shore. It made a shallow dive and swam back toward the beach where Whitbourne had come ashore, and his men were now rowing.

As the "mermaid" turned, Whitbourne noticed its back and shoulders were square, white, and "smooth as the backe of a man." From the middle to the end, it tapered to a tail that appeared to him to have the proportions of a "broad hooked arrow." The mystery animal headed toward another boat coming ashore. It placed its hands on the boat as if to climb aboard. One of the sailors hit it in the head with an oar, driving it back into the water. Later, it would approach two other boats, which would flee back to land.[1]

This account of grasping the boat's gunwale occurs repeatedly. This same behavior is chronicled in colonial New England and the Canadian Maritimes. Take, for example, this 1630s encounter in Casco Bay, Maine.

> One Mr. Mittin related of a Triton or Mereman which he saw in Cascobay, the Gentleman was a great Fouler, and used to go out with a small Boat or Canoe, and fetch a compass about a small Island, (there being many small Islands in the Bay) for the advantage of a shot, was encountered with a Triton, who laying his hands upon the side of the Canoe, had one of them chopt off with a Hatchet by Mr. Mittin, which was in all respects like the hand of a man, the Triton presently sunk, dying the water with his purple blood, and was no more seen.[2]

Are these attempts to capsize the vehicle or to simply climb in? Chapter Twelve will show that water apes also have a long history of terrorizing travelers by joining them on horses behind the ride. When technology advanced, the water apes climbed onto automobiles instead. There's no reason climbing aboard a small boat should be different. But is it a deliberate attempt to terrify, or are humans misinterpreting a playful act?

Consider this encounter in Canso Harbour, Nova Scotia, in 1656. A fishing vessel spots a merman cavorting in the water. Even when the crew attempts to net him, his escape seems more playful than panicked.

[The crew] saw clearly that this fish, or to say better, this monster, which still retained the same appearance, seemed to take pleasure in the beams of the sun (for it was about 2 p.m. and very clear and fine weather); it seemed to play in the gently undulating water, and looked somewhat like a human being. This caused general astonishment and likewise great curiosity to see this strange creature near by, and, if possible, to catch it … It came to pass that one of the sailors, or the fishermen, throwing out overboard away from the boat, cast a rope over the head of the Merman (for it was in fact a Merman), but since he did not quickly enough draw it shut, [the Merman] shot down through the loop and away under water, presenting in his lowest part, which because of the quick movement could not well be made out, the appearance of a great beast. At once all the boats gathered round in order to catch him in case he should come up once more, each one holding himself ready for that purpose with ropes and cords. But instead of showing himself there again above water, he came to view farther out to sea, and with his hands, whereof the fingers (if indeed the things were fingers that stood in the place of fingers) were firmly bound to each other with membranes just as those of swans' feet or geese feet, he brushed out of his eyes his mossy hair, with which he also seemed to be covered over the whole body as far as it was seen above water, in some places more, in others less. The fishermen distributed themselves again, and went a long way around, in order to make another attempt; but the Merman, apparently noticing that they had designs on him, shot under water, and after that did not show himself again, to the great dejection of the fishermen, who many a time went there to be on the lookout, and incessantly racked their brains to invent stratagems to catch him.[3]

Mermaid reports followed explorers and colonists, or vice versa. In 1676, the Royal Society ran an essay by Thomas Glover about Virginia in the publication *Philosophical Transactions*. The material emphasized that Glover had seen Virginia personally and that he was a "chirugion" (surgeon) who had lived in the country for multiple years. The Royal Society considered such a report on the climate and tobacco production by a colonist and professional man to carry more cachet. But Glover also had an odd addendum.

> And now it comes to my mind, I shall here insert an account of a very strange fish or rather a monster, which I happened to see in *Rapa-han-nock* River about a year before I came out of the Country; the manner of it was thus...I heard a great rushing and splashing of the water, which caused me suddenly to look up, and about half a stone's cast from me appeared a most prodigious Creature, resembling a man, only somewhat larger, standing right up in the water with his head, neck, shoulders, breast, and waste, to the cubits of his arms, above water; his skin was tawny, much like that of an *Indian*; the figure of his head was pyramid, and slick, without hair; his eyes large and black, and so were his eyebrows; his mouth very wide, with a broad, black streak on the upper lip, which turned upward at each end like mustachoes; his countenance was grim and terrible; his neck, shoulders, arms, breast and wast, were like unto the neck, arms, shoulder, breast and wast of a man; his hands if he had any, were underwater; he seemed to stand with his eyes fixed on me for some time, and afterward dived down, and a little after riseth at somewhat a farther distance, and turned his head toward me again, and immediately falleth a little under water, and swimmeth away so near the top of the water, that I could discern him throw out his arms, and gather them in as a man does when he swimmeth. At last he shoots with his head downward, by which means he cast his tayl above the water, which exactly resembled the tayl of a fish with a broad fane at the end of it.[4]

Glover was a passenger on a sloop ten miles upriver from Windmill Point, where the Rappahannock empties into the Chesapeake Bay. They were in the vicinity of the Towles Point bar, and they had mistimed the passage. The tide was ebbing, and the tidal waters were too low to risk crossing the bar. The sloop dropped anchor, and the sloopman and his mate took a small boat ashore to replenish their water supply, leaving Glover the sole witness.

Another of the earliest and most popularly cited reports came from Lake Superior in the 18th century. The eyewitness was a merchant and explorer known as Venant St. Germain. He made a sworn statement in 1812 to the event on May 3,1782 while he made camp on Isle Paté (Pie Island, near Fort William).

The account was given as a deposition before two judges of the Court of King's Bench for the District of Montreal. St. Germain had grown weary of being doubted and chose to tell his story under oath. It has become one of North America's most noticed historical reports of a merman.

A little before sunset, the evening being clear and serene, deponent was returning from setting his nets and reached his encampment a short time after the sun went down. That on disembarking, the deponent happened to turn towards the lake, when he observed, about an acre or three quarters of an acre distant from the bank where he stood, an animal in the water, which appeared to him to have the upper part of its body, above the waist, formed exactly like that of a human being. It had the half of its body out of the water, and the novelty of so extraordinary a spectacle excited his attention, and led him to examine it carefully. That the body of the animal seemed to him about the size of that of a child of seven or eight years of age, with one of its arms extended and elevated in the air. The hand appeared to be composed of fingers exactly similar to that of a man; and the right arm was kept in an extended position, while the left seem to rest upon the hip, but the deponent did not see the latter, it being kept under water. The deponent distinctly saw the features of the countenance, which bore an exact resemblance to those of the human face. The eyes were extremely brilliant; the nose small but handsomely shaped; the mouth proportionate to the rest of the face; the complexion of a brownish hue, somewhat similar to that of a young negro; the ears well formed, and corresponding to the other parts of the figure. He did not discover that the animal had any hair, but in the place of it he observed that wooly substance about an inch long, on top of the head, somewhat similar to that which grows on the heads of negroes. The animal looked the deponent in the face, with an aspect indicating uneasiness, but at the same time with a mixture of curiosity, and the deponent, along with three other men who were with him at the time, and an old Indian woman to whom he had given a passage in the canoe, attentively examined the animal for the space of three or four minutes.

The deponent formed the design of getting possession of the animal if possible, and for this purpose endeavored to get hold of his gun, which was loaded at the time, with the intention of shooting it; but the Indian woman, who was near at the time, ran up to the deponent, and seizing him by the clothes, by her violent struggles, prevented his taking aim. During the time which he was occupied in this, the animal sunk under water without changing its attitude, and disappearing, was seen no more.

The woman appeared highly indignant at the audacity of the deponent in offering to fire upon what she termed the God of the Water and Lakes; and vented her anger in bitter reproaches, saying they would all infallibly perish, for the God of the Waters would raise such a tempest as would dash them to pieces upon the rocks; saying that 'for her own part, she would fly the danger,' and proceeded to ascend the bank, which happened to be steep in that part. The deponent, despising her threats, remained quietly where he had fixed his encampment. That at about ten or eleven at night, they heard the dashing of the waves, accompanied with such a violent gale of wind, so as to render it necessary for them to drag their canoe higher up on the beach; and the deponent, accompanied by his men, was obliged to seek shelter from the violent storm, which continued for three days, unabated.[5]

On June 30, 1886, *The North Sydney Herald* reported that Gabarus Bay, Cape Breton Island fishermen had seen a mermaid in Nova Scotia. That newspaper did not survive in archives, but the other newspapers across Nova Scotia repeated their account.

The fishermen of Gabarus have been excited over the appearance of a mermaid seen in the waters by some fishermen a few days ago. While Mr. Bagnell, accompanied by several fishermen, were out in a boat they observed floating on the surface of the water, what they supposed to be a corpse. Approaching it for the purpose of taking it ashore for burial, they observed it to move, when to their great surprise, it turned around in a sitting position and looked at them and disappeared. A few minutes after it appeared on the surface and again looked toward them, after which it disappeared altogether. The face, head, shoulders and arms resembled those of a human being, but the lower extremities had the appearance of a fish. The back of the head was covered with long, dark hair resembling a horse's mane. The arms were exactly shaped like a human being's, excepting that the fingers on the hands were very long. The color of the skin was not unlike a human being. There is no doubt but the mysterious stranger is what is known as a mermaid, and the first one ever seen in Cape Breton waters.[6]

As shipping traffic increased and mills popped up on rivers, it disrupted the environment. It also brought more potential witnesses to the riverbanks. In 1881, it was the Susquehanna River in Pennsylvania.

One of our oldest fishermen reports the discovery of one of these rare nondescripts, in the river about one mile above town, in the deep water opposite Dugan's run. He has already seen it five times; always either early in the morning or late in the evening.

He says it comes to the surface, looks about it, then gradually sinks down leaving its hair floating on top of the water for a moment or so and finally disappears. It has the face of a woman and beautiful glossy black hair, but as it only shows itself down to the shoulders, he cannot tell what the other end is like.

He says he could shoot it but is afraid he might be arrested and tried for murder, and it would bring him into trouble. On being asked if it had a comb or looking glass with it. "It might have had, but he didn't see it" and supposes it has a cave somewhere in the bottom of the river under the deep water.

Mr. Henry Loucks, the fisherman above alluded to, is well known in town, and he is considered as reliable as any fisherman on the river. We are in hopes that the mermaid may be captured alive, if possible, or dead, if it cannot be had any other way, and guarantee him a safe delivery out of his troubles if he shoots it.[7]

In 1882, another article spread across the nation's newspapers, noting sightings of a creature in Big Charley Popka Creek, a tributary of the Manatee River in Tampa Bay.

A strange animal, half-man and half-fish, and covered with coarse, black hair, has been seen several times recently in Big Charley Popka River, Manatee County, Florida. The head and upper portion of the body resemble a man—short, stumpy arms, with webbed feet, taking the place of fins, while the lower portion is just like the tail of any fish. The monstrosity is about four feet in length, and when sporting about in the water, utters coarse, guttural sounds, a cross between the barking of a dog and the bellowing of a bull. Several attempts to capture and shoot the animal have proved futile.[8]

In 1894, the *Cincinnati Enquirer* reported on "creatures" seen in the Ohio River near Vevay, Indiana. The newspaper headline was "Nondescript are These Creatures, Which Resemble a Human Being in Many Ways." Most newspapers that carried the account went with the more succinct title of "Mud Mermaids."

From notes taken on the ground of the description as furnished by Mr Ozier states that the beast is about five feet in length and should weigh about 150 pounds. Its general color is yellowish. The body between the four legs resembles that of a human being. Back of the hind legs it tapers to a point. This point in no way resembles a tail. The legs, four in number, resemble the arms and legs of the human. The fore legs are shorter than the hind pair and are used in the same manner as arms. The extremities resemble hands and are webbed and furnished with sharp claws. On the back and one third of the way around the body appears a mass of straggling, coarse hair. The skin below the fore legs is thick and resembles elephant hide. On the arms and about the face and neck it is a finer texture and brighter yellow color than the rest of the body.[9]

In this 1894 account, we again see that not all merfolk present the same. Some appear with skins and tails, while others have hair. This suggests that water dwellers wear skins and appendages tailored to their environments. Alternately, perhaps different species of these creatures differ in appearance based on region. It appears they do not all need to do so.

The more modern accounts of merbeings across North America have tended to be brief encounters. There is a glimpse of something extraordinary and unexpected, leaving the eyewitness to marvel at what was seen. Such incidents breed legends, but they are not as informative as we would like them to be. There are several such glimpses, especially in states bordering the Great Lakes, where the Great Swamps of past centuries were once found. These extensive swamplands would have provided suitable habitats for merbeings. Today, only remnants of those wetlands remain. When we consider historical reports, especially those chronicling the unique tracks of North American mermen, we should assess what these reports can tell us about more recent sightings.

Something odd was in the woods around Charles Mill Lake near Mansfield, Ohio, in March of 1959. Two young men ran into something they initially thought was a log. It then stood up and walked towards them. It was seven feet tall and had green eyes. They fled the scene. Later tracks were found there that resembled swim fins.[10] In September 1963, there was a report in the *Mansfield News Journal* of a man spotting something in a nearby creek. Others in the area had also reported seeing a mysterious tall figure.[11] An extensive monster hunt ensued.

At three locations in Michigan, aquatic cryptids have been noticed, but only as word-of-mouth folklore. There has been talk about a merman seen in a swampy area at Sault Ste. Marie in Upper Michigan. Two fishermen saw something similar at Voorheis Lake in Oakland County in August of 1965. At Patterson Lake in Lower Michigan, there are stories of amphibious man-like creatures emerging from the lake.

Sightings of water creatures have also been reported in Arkansas. In Benton County, in the northwestern corner of the state, two women fishing in the Illinois River saw a four-foot-long creature in May of 1973 with green and black markings on its skin and long, black hair. Its eyes were big and yellow.[12]

For decades, there have also been discussions of such things in Conway, Arkansas. When Palarm Creek was dammed to make Lake Conway in 1950, the "monster" moved into the lake's shallow coastal waters.

> There have been reports of a huge monster splashing about in the creek, scaring fishermen and hunters with screams and squeals and kicking up quite a fuss in the water.
>
> About two weeks ago the rumors began circulating and growing in size and number...this time it was a weird creature playing havoc with things in general on Lake Conway. Reporter Walter Ed Scales of the *Conway Log Cabin Democrat* decided to trace down the source of the latest story.
>
> Scales finally contacted George Dylan of the Mayflower Community about 20 miles northeast of Little Rock. It seems Dylan was the angler who saw the creature.
>
> Here's the story as related by Dylan to Scales: Dylan was running a troutline one day when the thing floated to the surface as he raised the line. The creature stared at him a few seconds and then tried to shake the hook from its mouth.
>
> It had green spotted skin similar to that of a frog and a monkey like head. Its lips were blue, it had no teeth. Its hands were the size of a man's but it had webbed fingers and claws like nails.[13]

A watershed event in American culture took place in October of 1958. People across the United States became intrigued by the notion that Bigfoot was a large hairy hominoid afoot in the wilderness. People became aware of the presence of giant footprints and the possibility that such imprints were made by this legendary creature.[14]

In 1958, Bigfoot first stepped into the public consciousness. "Giant footprints puzzle residents," a headline in the *Humboldt Times* announced. The small Northern California newspaper reported that a road construction crew had discovered human-like footprints that were sixteen inches long. The memorable name "Bigfoot" originated from this time period and has been a term used in international cryptozoology ever since. The *Smithsonian Magazine* observed in 2018:

Today, the legendary beast seems to be everywhere: You will find Bigfoot looking awfully cute this year in two children's films: *The Son of Bigfoot* and *Smallfoot*. Animal Planet recently aired the finale of its popular series "Finding Bigfoot," which lasted 11 seasons despite never making good on the promise of its title. And the Bigfoot Field Researchers Organization lists at least one report from every state, except Hawaii, over the past two decades. The most recent sighting, in June 2018, was by a woman in Florida who reported a creature that looked like "a large pile of soggy grass." Other evidence in the database includes supposed Bigfoot scat, nests and noises. If a tree falls in the forest and no one is around to hear it, it may not make a sound—but it seems someone will report that a Bigfoot knocked it over.[15]

For years we have seen this interest grow. It has been fed by the fact that large footprints can be found all over North America. They are deposited by all manner of creatures, but many assume that they are left by Bigfoot.

One of these episodes took place in northern Alabama in August of 1978. It happened at a swimming and fishing spot called Blue Hole, located about five miles west of Athens, Alabama, in Limestone County. On the night of August 6, 1978, three teenagers from Athens were there on a trip to do some catfishing.

When the teens' campfire was dying down around 9:30 p.m., one of the campers drove his car alone down a dirt road to pick up more firewood. His headlights picked up some eyeshine, and the incident began.[16]

"I saw these pink eyes, and drove further into the woods. And I heard something growl. At first I thought it was just a deer or something. I got the headlights on it and it started to come towards the car. I threw the car in reverse and got out of there."

He drove back to his two friends. He shouted at them, "Come on. There's something up here and it's really big!" They thought he was joking. He convinced them to go along, and they drove back down the road. The boys said that even though the engine was loud, they could hear something trampling through the woods. Another one of the teens reported:

I was revving the engine up on my car so the lights would be better. My battery is kinda low so I was doing this to get the lights up. Suddenly, the creature broke out of the brush on my side of the car. It was about eight feet from the car. I shoved it in reverse and headed back to town 90 or nothing.

These particulars were given: "The boys described the creature as very big, eight or nine feet tall, blackish-brown in color with glowing pink eyes; with long hanging arms; making a breathing kind of growl and possessing a peculiar smell. They said Stanley Estep commented that they had noticed the same smell earlier but had not paid any attention to it." David Conley added, "The fur on the creature was very coarse and looked like steel wool in the car light." They thought it was man-like but bigger.

Finding footprints with five toes in the same area is not unique. The tracks of merbeings have been reported in the Big Thicket of Texas, the same area as the five-toed tracks. Such tracks suggest that the activities of living fossils overlap in these areas.

Water apes have also been seen along the West Coast as well. A 1935 sighting was recorded by Captain Sam Orlando and the eleven crewmen of the 65-foot purse seiner *M.K. No.1*. This encounter includes an earlier report of a possible infant.

"We had a hard night, and only the helmsman. Frank Verga, was on deck. Frank thought he saw a runaway buoy, and there's a reward for catching one, so he steered for it. As he got closer, he let out a yell, and we all ran up.

There it was, with its head a couple feet out of water, looking straight at us, with its shiny eyes under a broad smooth forehead. It had brownish gray hair, two or three inches long on its head and under its chin. I guess you could say it had whiskers.

How big? Oh, 10 or 12 feet long. We watched it swim across our bow, and Frank turned off the engine. There it stood still staring. It made some of the boys shake their heads but we only had about six tons of fish aboard, so I ordered the net out and the skiff manned.

The cannery hadn't posted any price on a bearded merman, but I figured we might sell him to a pleasure pier concession. But we made so much noise about it all, that he dove under, flipping his tail at us as he disappeared."

Captain Orlando and another member of his crew, Mateo Giacolone, said they saw a similar creature on August 17. 1922. What they saw 13 years ago, they said, apparently was a mermaid, or possibly a mermatron, as it was holding, Captain Orlando swore, and still does, a merbaby in its flippers.[17]

In August of 1972, campers reported seeing something emerge from the water at Thetis Lake on Vancouver Island in British Columbia. It made the news, and The Royal Canadian Mounted Police investigated the reports. Two other teenagers reported seeing a creature emerge from the water, look around, and then go back under the surface. They described the being as human-shaped with a monster face. It was silver, covered in scales, and had big ears with a point sticking out of its head.

The notable thing is the range and timespan of these accounts. As explorers and settlements moved deeper into North America, they found they were not alone.

ENDNOTES

1 Whitbourne, "A conclusion to the Reader,"*A Discovrse and Discovery of Nevv-Fovnd-Land* (1620).
2 Josselyn, *An Account of Two Voyages to New-England* (1674), 23.
3 Denys, *Description & Natural History of the Coasts of North America (Acadia)*, (1908), 80–81.
4 Glover, "Account of Virginia, *Philosophical Transactions* 11(20), June 1676.
5 "A Mermaid in Lake Superior," *The Canadian Magazine* 11(2), May 1824.
6 "A Sea Nymph Seen at Garbarus," *Saint John Colonist,* 9 July 1886.
7 "A Mermaid in the Susquehanna," *The York Daily,* 8 June 1881.
8 "A Strange Animal in Florida," *Brooklyn Daily Eagle,* 1 October 1882.
9 "Nondescript are These Creatures," *Cincinnati Enquirer,* 6 September 1894.
10 "Boys Report Seeing Green-Eyed Monster," *Mansfield News Journal, 28 March 1959.*
11 Gaynor, "'Hairy Monster' Has Folks in a Tizzy," *Mansfield News Journal,* 27 July 1963.
12 "Little Monster Breaks Fishing Poles of Women," *Northwest Arkansas Times,* 28 May 1973.
13 "Arkansas Creek Monster Tale Again Appears in Conway Area," *The Courier News,* 11 March 1952.
14 Hall, "October 1958 in the History of Bigfoot," *Wonders* 9(3), September 2005.
15 Crair, "Call of the Wild Man," *Smithsonian* 49(5), September 2018.
16 Sutton, "Fishing trip results in sighting of 'Big Foot'?" *News Courier,* 8 August 1978.
17 "Merman Seen by Fish Boat Crew off Port." *San Pedro News-Pilot,* 25 May 1935.

Chapter Five
South America and the Caribbean

EUROPEAN EXPLORERS BROUGHT their traditional version of the half-woman, half-fish mermaid to South America. According to Persephone Braham, professor of Spanish and Latin American literature at the University of Delaware, any native stories of aquatic beings were shoehorned into the colonist preconceived version of mermaids, although Braham questions how much the European version was absorbed into local traditions. Braham further notes, "from Patagonia to the Caribbean, the social and moral function of Latin American mermaids is fundamentally different from that of their European counterparts."[1]

The Caribbean has the New World's oldest account of a merbeing filtered through European bias. In 1493, Columbus was exploring the Caribbean, following the Hispaniola coast (Haiti) to the Yaque del Norte river, when he saw three mermaids rising out of the waves. Although they had some human appearance, they were not as attractive as he expected, with masculine traits on their faces. He also noted he had also seen mermaids in the African waters off Guinea. Although a manatee is unlikely, considering these seasoned sailors were familiar with the species, a water ape would likely generate similar disappointment to a European expecting a buxom, blonde fish-woman.

South American merfolk reports date back earlier than North America. This is because the explorers and colonists brought Jesuit priests as missionaries and chaplains. These men of God would return to the Old World and publish their New World impressions to an eager audience. These Jesuit missionaries said the New World was filled with unusual creatures.

Unfortunately, these missionaries were not trained naturalists, so the accounts are often confusing and conflicting.

The pan-African mermaid/water spirit *Mami Wata* was brought to the new world with enslaved Africans, according to the research of Roger Bastide, a French sociologist and anthropologist who specialized in Brazilian literature. After this, the tales of the spirit evolved as the goddess was melded with Catholic saints. The new gods and goddesses became hybrids of African, South American, and religious figures, superficially Catholic enough to keep the missionaries content but still distinct.[2]

Professor Bettina E. Schmidt, professor of religious studies and anthropology, notes that this confusion continues.[3] She studied the Brazilian water goddesses *Oxum* and *Iemanjá*, two deities of the Candomblé religion (one of several religions with African roots). Schmidt notes that the mermaid may be homogenized across the nation, much as *Mami Wata* became a singular figure across Africa.

Schmidt discovered that *Oxum* is a freshwater mermaid figure. *Iemanjá* was a freshwater figure now considered a saltwater goddess, which makes more sense to her since "Mermaids are usually connected to a saltwater environment."[4] But, this is the European bias showing. The Haitian merbeing/water goddess *Lasirén* is another version of the European-influenced water figure also associated with seduction and wealth. The water apes are equally at home in saltwater as in freshwater. In Chapter Seven, we'll discuss why.

Perhaps it is best to search Latin America and the Caribbean for our aquatic cousins using a different word than "mermaid," which has so many preconceptions attached to it.

Trinidad and Tobago were colonized by the Spanish in 1498. In 1797, they were British, but as war and maritime dominancy ebbed and flowed, the islands were also Dutch and French territories. The Afro-Caribbean descendants of Natives brought to the islands by the French and the Indo-Caribbean descendants of British indentured workers remain the two largest ethnic groups. The mermaid representations of the pan-African *Mami Wata* are not half-woman, half-fish. On these islands, mermaids have lost their fish half in favor of a serpentine lower half. This could be the influence of the East Indian residents. And as mermaid folklore shows,[5] there are more than forty species of snakes on the islands, including the green anaconda, one of the largest snakes in the world. A water ape could adapt to available resources, and snakeskin might be logical and plentiful for protective gear.

French folklorist Michel Meurger, who has extensively studied lake monsters, says the *hipupiára*, also called the *ipupiára* or *igpupiára*, was a generic catch-all term for unknown animals.[6] The encounters with the *hipupiára* accounts sound like water apemen. The most famous record of an encounter with a *hipupiára* was a battle. It was eleven feet long and covered with hair. The combat was sufficiently riveting that Portuguese historian Pero de Magalhães Gandavo devoted an entire chapter to the incident in his 1576 history.[7] In 1554, in the Portuguese Captaincy of São Vicente, Baltesar Ferreira, the son of the colony's leader, was awakened by the screams of his father's Native slave. She had been outside and seen a monster on the beach that, to her, must have been Satan himself. Ferreira proceeded to the beach, and to his shock, there was indeed a monster.

The monster saw Ferreira and started back to the water. Ferreira intercepted and barred its path. The beast stood upright like a man. Ferreira thrust his sword into the beast's belly. The resulting geyser nearly blinded Ferreira as the beast attacked with teeth and claws. Ferreira was able to cut the beast again in the head. Weakening from blood loss, the beast turned again toward the ocean. Other slaves had arrived, seized the nearly dead monster, and paraded it around the town.

In the 1580s, the male *Igpupiára* was described as a traditional European creature, looking reasonably tall with very deep-set eyes. The female had the appearance of a beautiful woman with long hair.[8]

In 1578, French protestant Jean de Léry recorded an account from the southeastern Brazil colony of Villegaignon in Guanabara Bay. A Native was fishing at sea when an *Igupiára* surfaced and put its fin in the boat. The terrified Native chopped off the appendage and found it had five fingers, like a hand. The Native also said the *Igupiára* had a human face.[9] This event is also notable because it reappears in accounts from North America. Léry's version, repeated in slightly differing forms over the centuries (most notably

The Ipupiara, according to the 16th century Tupis of Brazil, was a sea monster and anthropophagous. Ipupiára by Pêro de Magalhães Gândavo. *Wikimedia Commons.*

the Casco Bay account in Chapter Four), is either plagiarism or indicates a water ape behavior that involves grabbing a boat. If the latter, it may be related to Lizardmen jumping on horses behind the rider. We'll discuss that in a later chapter.

In 1602, Martín del Barco Centenera, the chaplain of the Juan Ortiz de Zárate expedition to Rio de La Plata, completed an epic poem. It is of little value as art but invaluable as a historical record of Spanish activity in Argentina. Barco Centenera notes the presence in the river of *peces de humana forma*, fishes much like humans. He later specifically refers to them as "*de la 'irena*" or mermaids in a description of the flora and fauna of the region.[10]

In the 1550s, as the Spaniards founded the city of Coquimbo, a mermaid was spotted off the coast. It was not a rare occurrence. In fact, it was believed that the neighboring city of La Serena was named after the mermaids, according to Spanish chronicler and author Diego de Rosales. In his 1674 book *Historia General del Reino de Chile* (*General History of the Kingdom of Chile*), he notes that the *sirena* in Spanish is referred to by the Natives as *pincoy*. He adds that in 1632, a large group of colonists and Natives saw a mermaid approach the beach. She had the face and breasts of a woman with flowing blonde hair. She carried a child in her arms. She had the tail and back of a fish when she dove, visibly covered with thick scales "like little shells." In the book, instead of remarking on the unusual frequency of mermaid spotting, de Rosales complains that there are no reports of mermaids singing because everyone in Europe knows they sing![11]

At the end of June in 1794, the British merchant ship *Rattler* was sailing south along Chile's coast under Captain James Colnett. He had sailed to the Galapagos Islands on behalf of the Admiralty and private whaling interests. Colnett was sent to assess the potential of developing whaling as an industry near the islands. Having mapped the islands and loaded the ship with oil from encountered sperm whales, the ship was heading for Cape Horn. The crew was uneasy despite the profit from the whale oil in the holds. Colnett hated the Spanish, and South America was Spanish territory. There were concerns that an encounter with a Spanish vessel would end badly. Once while fur trading with the Spanish in the Pacific Northwest, Colnett saw his ship seized and the crew put in irons. The longer they traveled along the South American coast, the more nervous the crew became. Little things became omens of ill fortune. Then things took a decidedly odd turn.

When we were in Latitude 24°, a very singular circumstance happened, which as it spread some alarm among my people, and awakened their superstitious apprehensions, I shall beg leave to mention. About eight o'clock in the evening an animal rose along-side the ship, and uttered such shrieks and tones of lamentation so like those produced by the female human voice, when expressing the deepest distress, as to occasion no small degree of alarm among those who first heard it. These cries continued for upwards of three hours, and seemed to increase as the ship sailed from it.[12]

Colnett suggested to the crew that a female seal had lost its cub. But even he had to admit he had "never heard any noise whatever that approached so near those sounds which proceed from the organs of utterance in the human species." The crew considered this as another evil omen, and Colnett was forced to worry whether the seamen were becoming too despondent to rally should the need arise.

In 1797, British physician Colin Chisholm was on a social call with Abraham Jacob van Imbyze van Batenburg, the governor of Berbice (now Guyana). The governor told Chisholm of the mermaids found in the Berbice rivers, called the *méné mamma*, or "mother of the water" by the local Natives. The governor described the mermaids as follows.

The upper portion resembles the human figure, the head smaller in proportion, sometimes bare, but oftener covered in copious quantity of black long hair. The shoulders are broad, and the breasts large and well formed. The lower portion resembles the tail portion of a fish, is of immense dimension, the tail forked, and not unlike that of a dolphin, as it is usually represented. The colour of the skin is either black or tawney. The animal is held in veneration and dread by the Indians, who imagine that the killing it would be attended with the most calamitous consequences. It is from the circumstance that none of these animals have been shot, and consequently not examined but at a distance. They have been generally observed in a sitting posture in the water, none of the lower extremity being discovered until they are disturbed; when by plunging, the tail appears, and agitates the water to a considerable distance.[13]

South American reports of the fish-man have also been recorded in Paraguay, where relations between the local peoples and their aquatic cousins appear less than cordial. A missionary recorded the tale in his 1914 book that only came to the attention of folklorists thirty years later when it was reprinted by the American Folklore Society.

> In a certain Chaco river, a monster was supposed to live, who had the form of half-man and half-fish. He was held in great terror by the people, and few dared to approach the river. A woman one day unthinkingly went down to draw water, and met the fish-man, who fell in love with her, and asked her to become his wife. She, out of fear, consented, but begged leave to first return to the village with her water-jar. Her request was granted.
>
> For a long time she was afraid to return to the river, but eventually she had to go, as there was no other water obtainable. As she was dipping in her jar, the fish-man again appeared, and was very angry with her for having deceived him. Seizing her by the hand, he insisted on taking her to his home beneath the water, and making her his wife. She, however, bit his hand, and, as he drew it away in pain, she ran off. In revenge he caused the water of the river to flood, with the result that nearly all her people were drowned.[14]

On the islands of the Chiloé Archipelago, there's a difference between mermaids and the *pincoya*. Pincoya and her husband, Pincoy, are also aquatic but do not have fishtails. She appears completely human. Her husband has a human face and the body of a large seal with silver fur.[15]

In Brazil, the guardians of fish and other underwater denizens were the *Uauyará*. They could transform into the freshwater pink river dolphin that lived in the river. In a historical account from 1876, it says that in Pará, the northern Brazilian state crossed by the lower Amazon River, there wasn't a single village in the interior that did not have its own stories of the *uauyará*. There's even a rare account of the *uauyará* gathering together in the rainforest to sing songs and play drums.[16]

The Moche civilization of northern Peru was noted for advanced agriculture and art used to share legend and history mnemonically. Ai Apaec, the hero ancestor of the Mochica civilization, battled the sea's most powerful deity/demon. The Moche people had no writing system, so scholars nicknamed this aquatic foe "The Decapitator." The Decapita-tor is depicted as a hybrid of different creatures, usually part shark, part ray, and part sea lion. This sounds similar to a protective suit of an aquat-ic ape cobbled together from available skins. The Decapitator's bloody reputation could be interpreted as a battle for water resources between apes and Andeans.[17]

The April 22–29, 1736, issue of the *Pennsylvania Gazette* reported a merfolk sighting. The brief article was taken from an unnamed newspaper in Bermuda and reported that a sea monster had been spotted recently. The upper body appeared to be a 12-year-old boy with long black hair and the lower half of a fish. He was first spotted on the shore and fled to the water with a boat in hot pursuit. Just as the pursuers were about to drive a fishgig into him, the unexpected human likeness shocked the crew into compassion, and they could not kill the creature.[18]

Frederick Carruthers was a successful real estate man in New York City interviewed by the *Brooklyn Daily Eagle*. Carruthers was a former sailor, and one of his stories had reached the desk of *Eagle* editor Charles Skinner, a successful folklore collector. Carruthers recalled standing on the deck of an anchored barque off Demerara in British Guiana (present-day Guyana). He was facing the bow, taking a sponge bath. As he drew a bucket to douse him-self, he observed a dark-skinned mermaid on a rock about thirty feet away:

> "The face was quite round, a perfect human face of small size, and the hair hung far down the back [shoul-der-length]. The color was a little darker than my hand [a deeply tanned Caucasian].
> The neck was human, and the shoulders somewhat resembled the shoulders of a woman, but the body rapidly became fish, with big fins and a regular fish tail.[19]

Carruthers called to the cabin boy, but his voice alerted the creature, and it disappeared. By the time the cabin boy arrived, the mermaid was gone. The unnamed reporter may not have understood his editor's inter-est as a folklorist. Instead, the article was light-hearted, gently mocking Carruthers for stereotypical sailor tall tales. Other newspapers carried the story, cut the original piece in half, and removed the ribbing that diluted the story.

Some of these traditions have been very clear about the presence of merbeings where people say they have known them for generations.

In the Caribbean, we encounter some classic elements of the interaction between mermaids and men; not all are centuries old. For example, one newspaper correspondent reported from George Town on the island of Exuma.

> According to folklore, mermaids and mermen were thought to belong to the natural order although having certain characteristics suggesting kinship with the supernatural. They loved music and were often heard singing. On Stocking Island the sound of surf mingling with the sighs of the trade winds in the palm fronds can easily be mistaken for a siren's song.
>
> Numerous folk tales record marriages between man and mermaid. In legend, the man first steals the mermaid's cap or belt, her comb or mirror, and thus gains power over her. As long as these objects remain hidden, she lives with him. But if she finds them she returns at once to the sea.
>
> Belief in the existence of these hybrid people is curiously persistent, claims of having seen them or heard them have been made in comparatively recent times. It has been said that mermaids sometimes lured mortals to death by drowning, or enticed young people to live with them under water. Pools, rocks, and caves are their special haunts, and a grotto on Stocking Island is believed by many people to be inhabited by a particularly alluring mermaid.
>
> Exploring visitors, unaware of the legend surrounding it, might be even more intrigued to learn that any man daring to swim in the grotto's cool depths risks confrontation with the supernatural. Legend has it that a beautiful mermaid will surface from the cavern's waters and pull him down to her domain."[20]

In the Bahamas, teenager Samuel Lowe had a skiff (a shallow, flat-bottomed open boat used for coastal fishing propelled by pushing against the sea bottom with a pole). He recalled poling with a friend to a place known locally as "Conch Rocks," a small grouping of eight-foot rocks in a dangerously shallow area several miles off Clifton Bluff on the southeast shore of New Providence Island. Because the waters were shallow enough to be a navigational hazard to larger ships, it meant good fishing for a small skiff. Lowe recalled the day was clear with no clouds in the blue sky, and they could see for miles.

Conch Rocks is on the edge of the "tongue of the ocean," a region of deep water in the Bahamas separating the islands of Andros and New Providence. As the teens came into view of Conch Rocks, they noticed something on one of the taller rocks near the drop-off into the depths. They knew this was a dangerous area for a small boat. They positioned the skiff within twenty feet of the rocks to get a better look. It was a woman, miles away from dry land and with no signs of wreckage. Lowe's friend was convinced it was a corpse washed up on the rock.

Lowe wasn't sure who or what she was, but he thought she was sitting waist-deep in the water on the rock and moving. Despite increasing nervousness, they poled the craft in a little closer. Her hair was green-black and looked like seaweed. She threw her arms over her head, and Lowe was fascinated to see her arms were flippers like a turtle. She had no hands or fingers. The creature ignored the boys as they shouted and waved. She continued to gaze seaward. Lowe wanted to get closer, but his friend was hysterical with fear. He remained convinced it was a corpse based on the unnatural color of the being.

Lowe had misgivings about leaving the potentially stranded woman behind. But he had the clear impression that she had no interest in being rescued or chatting with a pair of dumbstruck adolescents. That evening, when Lowe told the story over dinner, his parents, descendants of sea-faring people, suggested that the woman was a mermaid and suggested Sam avoid that area in the future.[21]

In 1968, a diver named Robert Froster was diving in search of stone structures off Bimini Island. At the last minute, his dive buddy canceled, but Froster, breaking a cardinal rule, decided to dive alone. As he swam, he noticed something in his peripheral vision. Something was stirring up the bottom sediment. He saw something heading toward him in the now murky water with an undulating motion. When it got within twenty yards, the Australian diver noticed that the creature had arms that appeared to be tipped in talons, and it was reaching out to him. As it drew closer, Froster saw it was a mermaid. Her top half had skin, breasts, and a head of hair, while the other half was a fish. She did not appear friendly, and the increasingly frightened diver would recall her gaze as nothing short of hateful. Froster chose discretion over science, raced to the surface, and boarded his vessel without further incident. Froster would later recall that he felt the mermaid had planned to ambush him, and he would be dead if he hadn't noticed the sediment being stirred up.[22]

Another storied place where mermaids are known to frequent is outside the town of Jacmel in southern Haiti.

> Popular with Jacmel visitors is a 90-minute horseback ride to the three Basins Bleu (Blue Lakes) in the nearby hills. According to legend, water nymphs inhabit the three mountain grottoes, and the goddess of waters sits on a famous rock beside the Palm Lake combing her hair by the light of the moon. When mortals appear, she disappears into the waters, and anyone who finds her comb, the story goes, becomes instantly wealthy.[23]

In Jamaica, Bog Walk Gorge in St. Catherine has a bridge across the Rio Cobre River with a bad reputation. The infamous "Flat Bridge" is only wide enough to accommodate a single traffic lane. It has no side barriers and is so low that it often floods. It looks shallow, but many vehicles have been recovered from the water, proving otherwise. Government officials blame the deaths on heavy traffic and impatient drivers. The residents believe it is River Mumma, a particularly blood-thirsty "very hairy half-human, half-fish." Daniel Gayle, a local diver and fisherman, says there is a season when the mermaid wants blood. And that is when accidents usually occur. In an article in the *Weekend Star*, a man named Samuel Dixon recalls a fisherman named Roaster. In 2010, Roaster was showing off a large fish scale to Dixon. Roaster claimed he shot it off River Mumma. Dixon recalled, "Roaster decided that him want to go back to see this fish with tail and human head and him never make it back. We find him drown with him neck and hand bruk."[24]

When two fatal crashes occurred a week apart in 2021, the newspaper's print edition downplayed any local folklore.[25] However, the online version went full tabloid, emphasizing the "supernatural" history of the Rio Cobre, not the fatalities.[26] It is unfortunate for the deceased, but rich in local lore for the cryptozoologists.

There are also traditions of water apemen in Central America. Eduard Conzemius came upon them when making an ethnographic study of the Miskito and Sumu people in Honduras and Nicaragua. They live along the Atlantic Ocean between Rio Tinto (Black River) and Rio Punta Gorda.

The mermaid or water nymph is an evil water animal, which occasionally causes snags and strong ripples where the water otherwise is very smooth. It drives the fish away so that the Indian can not catch anything, and it incites the alligator to attack the canoes and upset them. It also assumes the shape of a beautiful woman and walks on land to entice the young men down to the waterside, when suddenly it pushes its victim into the Water and devours it. Its head is that of a human being, but the body resembles that of a fish. This monster is also said to inhabit the sea, where it occasions waterspouts and hurricanes; the Miskito living at the coast call it also kabo wlaska "sea spirit."[27]

ENDNOTES

1 Braham, "Song of the Sirenas" in *Scaled for Success* (2018), 150.
2 Bastide, *The African Religions of Brazil* (1978), 259.
3 Schmidt, "Mermaids in Brazil" in *Anthropology and Cryptozoology* (2016), 157–170.
4 Schmidt, 161.
5 Bacchilega, Cristina, and Marie Alohalani Brown, *The Penguin Book of Mermaids* (2019), 273.
6 Meurger, *Lake Monster Traditions* (1988), 200.
7 Gândavo, *História da Província Santa Cruz* (2008), 53.
8 Cardim, *Tratados da Terra e Gente do Brasil* (1925), 89.
9 Léry, *Histoire d'un Voyage fait en la Terre du Brésil* (1578), 190–191.
10 Barco Centenera, *Argentina y Conquista del Rio de la Plata* (1912), II; 17.
11 Rosales, *Historia General de el Reyno de Chile* (1877), 309.
12 Colnett, *A Voyage to the South Atlantic and round Cape Horn into the Pacific Ocean* (1798), 169.
13 Chisholm, *An Essay on the Malignant Pestilential Fever* (1801), 193.
14 Grubb, *A Church in the Wild* (1914), 60–61. Métraux, *Myths of the Toba and Pilagá Indians of the Gran Chaco* (1946), 30-31.
15 Cavada, *Chiloé y los Chilotes* (1914), 101–103.
16 Couto de Magalhães, *O Selvagem* (1876), part II 137–8.
17 Holmquist and Frareso, *Machu Picchu and the Golden Empires of Peru* (2021), 132.
18 "Philadelphia, April 29." *The Pennsylvania Gazette*, April 22-29, 1736.
19 "He Did See a Mermaid." *Brooklyn Daily Eagle*, October 7, 1894.
20 "Mythical maidens come to life in the Bahamas," *Winnipeg Free Press*, 1 March 1975.
21 Cappick, "Key West Resident Sees a Mermaid," *Paths*, July 1934.
22 Goudsward, *Sun, Sand, and Sea Serpents* (2020), 221–222.
23 Clarke, "New roads show Haiti's side not seen in cities," *Chicago Tribune*, 13 February 1977.
24 "Bog Walk gorge is Haunted," *The Weekend Star*, 15 December 2017.
25 Mathison, "Woman killed in Bog Walk Gorge crash," *The Gleaner* 9 June 2021.
26 Mathison, "Flat Bridge haunted—Residents claim mermaid lives in Rio Cobre," *The Star*, 9 June 2021.
27 Conzemius, *Ethnographical Survey of the Miskito and Sumu Indians of Honduras and Nicaragua* (1932), 167.

Chapter Six
Europe

WE HAVE SEEN THAT TRADITIONS of merbeings can be found throughout the Americas, as biologist Richard Carrington notes in the book *Mermaids and Mastodons*.

> There is not an age, and hardly a country in the world, whose folklore does not contain some reference to mermaids or to mermaid-like creatures. They have been alleged to appear in a hundred different places, ranging from the mist-covered shores of Norway and Newfoundland to the palm-studded islands of the tropic seas.[1]

The reasons for such a diversity of creatures and locations are due to the origins of *Oreopithecus*. They have adapted to survive in many climates. They have had the time to evolve their own physical and cultural means of survival in the many conditions found in the aquatic world. Their success was nearly universal, judging by the historical records.

There are accounts of them appearing in the polar regions—areas so remote that only people who are interested in polar exploration have trod them. One of those people was Henry Hudson, captain of the *Hopewell*. On his second voyage, he had just entered the Barents Sea on June 15, 1608, when two crewmen reported a mermaid sighting.

> This morning, one of our companie looking ouer boord saw a mermaid, and calling up some of the companie to see her, one more came up, and by that time she was

come close to the ship's side, looking earnestly on the men: a little after, a Sea came and ouerturned her: from the Nauill upward, her backe and breasts were like a womans, (as they say that saw her) her body as big as one of us; her skin very white; and long haire hanging downe behinde, of colour blacke: in her going downe they saw her tayle, which was like the tayle of a Porposse, and speckled like a Macrell.[2]

At the edge of Antarctica, Captain James Weddell told how a sailor made a similar report at the opposite end of the world.

> The sailor had gone to bed, and about 10 o'clock he heard a noise resembling human cries, and as daylight, in these latitudes, never disappears at this season, he rose, and looked around, but on seeing no person, he returned to bed; presently he heard the noise again, and rose a second time, but still saw nothing. Conceiving however, the possibility of a boat being upset, and that some of the crew might be clinging to some detached rocks, he walked along the beach a few steps, and heard the noise more distinctly, but in a musical strain.
>
> On searching around, he saw an object lying on a rock, a dozen yards from the shore, at which he was somewhat frightened. The face and shoulders appeared of human form, and of a reddish colour; over the shoulders hung long green hair; the tail resembled that of the seal, but the extremities of the arms he could not see distinctly. The creature continued to make a musical noise while he gazed about two minutes, and on perceiving him it disappeared in an instant.[3]

The most telling aspect of these reports is evidence that merbeings have successfully spread to the inhospitable areas near the poles. This adds to the impression that the creatures have been successful worldwide and have great longevity as part of the primate world.

Despite the widespread accounts, the lack of recent fossil records is accountable for three main reasons. First, they are water-dwellers whose remains would be readily destroyed by natural forces. Second, they appear to be cultural beings who would decide their own means of disposing of

Mermaids swimming around a galleon from *Aeneis Virgiliana* by Publius Vergilius Maro, 1529.

their dead comrades, and we don't know what those means are. And third, we have made no effort to seek the remains of water-dwelling primates in recent fossil deposits where they might have been preserved by chance. This last effort is usually essential in successful fossil hunting; a specific effort is made with a goal in mind.

Another crucial step in building fossil records is to correctly organize what we have turned up. First, one must recognize the fossils found for what they indeed are. Some fossil finds languished until their true nature was recognized. There is no end to accounts of fossils identified correctly only after years of languishing in museum vaults.

It cannot be said often enough that the differences in descriptions are not a valid reason to dismiss these findings. Instead, they may suggest a variety of adaptations (someday to be recognized in terms of genera and species). They may also reflect the cultures adopted by water apes in different localities.

Roman scholar Pliny (AD 23–79) claimed that mermaids are natural, not fabulous. In his *History of the World,* they were included. Here is what he had to say.

As for the meremaids known as Nereides, it is no fabulous tale that goeth of them: for look how painters draw
them, so they are indeed; only their bodie is rough and
skaled all over, even in those parts where they resemble a
woman. For such a Meremaid was seen and beheld plainely
upon the same coast neer to the shore: and the inhabitants
dwelling neer, heard it a farre off when it was dying, to make
pitteous mone, crying and chattering very heavily. Moreover,
a lieutenant or governour under *Augusus Cæsar* in Gaul, ad-
vertised him by his letters, That many of these Nereides or
Meremaids were seen cast upon the sands, and lying dead.
I am able to bring forth for mine authors divers knights of
Rome, right worshipfull persons of good credite, who testify
that in the coas of the Spanish ocean neerr unto Gades, they
have seene a mere-man, in every respect resembling a man
as perfectly in all parts of the bodie as might bee. And they
report moreover, in the night season he would come out of
the sea abourd their ships: but look upon what partsoever
he setled, he waied the same downe, and if he rested and
continued there any long time, he would sinke it cleane.[4]

This sort of accidental encounter is common in records. One of the
better-known reports of a legged merman comes from Ralph of Coggeshall, found in an 1197 account in his *English Chronicle*.

When King Henry II was guarding Bartholomew de
Glanvill's castle at Oreford, it happened that the fishermen
who were fishing there in the sea, caught a woody man
within their nets; who, being delivered over to the aforesaid
castle by reason of admiration, was in every respect naked,
and displayed the human form in all his members. He had
hair, however, on the surface, as it were mangled and torn;
but his beard was long and covered with hair, and around
his breast too hairy.[5]

It appears that such encounters were hard to avoid. In 1215, there
were reports of a large number of merfolk dwelling in the seas around Britain. In 1250, a Norwegian author complained the waters off Greenland
were "infested" with merbeings.[6]

England continued to have accidental encounters. An excellent example of this continuing issue is a mermaid snared in a salmon net in 1838. The story found its way into the American newspapers.

> This is certainly a "sea woman" and has such an expression of intelligence in its countenance, that we are absolutely inclined to believe that it is a creature of reason, rather than of instinct. We do not mean to insinuate that it has any notion of abstract ideas, but the fact is, that there is an appearance of confusion about it, that would seem to indicate sentiments of shame, and supplication.
>
> It weighs about seventy pounds and is altogether human in its outward organization from the head to the navel, when the fins begin to develop itself, and the remainder is formed very like the extremity of a large dolphin.
>
> It looks to be twelve or fourteen years old, and regards people, occasionally, as if it had an inclination to speak; and we are solemnly of the opinion that, when in its native element, it makes its wishes known through the medium of its tongue. Altogether, indeed, it is the most singular being we have ever witnessed, and excites feelings in the breasts of beholders, at least as much akin to awe as to curiosity. Can it have a soul, and be an accountable creature?[7]

Merbeing folklore in Great Britain is a fascinating study. It is a blend of different cultures with distinctive versions. Some have tails, some have legs, and all have different behaviors and aggressiveness. It makes sense if you assume communities of water apes adopted differing styles of protective garb and their interaction with the local population is based less on the location and more on their treatment by humans. For instance, merfolk along the Welsh Border area do not live in the sea but in freshwater lakes and rivers. They also seem to be curious about human artifacts. In Marden, Herefordshire, a church bell once fell into a deep pool in the river, and a mermaid took it away before the parish could find a way to retrieve it.[8]

English folklorist Jacqueline Simpson believes some mermaid accounts came to England with the Vikings.[9] Both Scandinavian and British water-dwellers notably lack fishlike tails. Is this evidence of Norse folklore becoming British, or was a Scandinavian water ape culture expanding into Great Britain before the Norse?

In Celtic folklore, merbeings who lose their hat or cloak cannot return to the sea, much like divers who can't submerge without their headgear. The "Man who Married the Mermaid" tale exists in several hundreds of versions from nearly all coastal counties in Ireland, making it one of the more popular stories.[10] In the story, a man meets a mermaid and learns that she can't return to the ocean if he takes away her comb. So he steals it from her to keep her on land, then marries her. The important detail in this story, often overlooked by folklorists, is that the tale teaches how merwomen combed their hair. With their heads bent forward and hair hanging down in front, it obscured their vision and made it easier for a man to steal her property. This also could include obstructing a water ape's vision while removing a protective hood. So, it can be interpreted as a folktale or a tactical maneuver if the water apes and humans have issues. And there is no shortage of accounts of merbeings coming into conflict with humans.

A mermaid was spotted on the rocks along Ireland's Mannin Bay near Derrygimla in 1819. The creature was startled and headed toward the water only to discover the tide had gone out, leaving the mermaid struggling to get to the water. Once in the safety of the water, she was studied by the large crowd that had gathered.

> The size of a well grown child of ten years of age; a bosom promiment as a girl of sixteen, a profusion of long dark brown hair; full dark eyes, hands and arms formed like the human species, with a slight web connecting the upper part of the fingers, which were employed in throwing back her flowing locks, and running them through her hair; her movements in the water seemed principally directed by the finny extremity.[11]

Witnesses said the mermaid was "half female and half fish, her lower extremities having the conformation of a dolphin." She watched the crowd for nearly an hour, only vanishing after someone fired a musket at her.

The vengeful nature of merfolk in French traditions is exemplified by an account noted at La Repentie near La Rochelle. A wealthy couple with several children paid a terrible price for capturing a mermaid. The couple were out fishing and caught the maid in their nets. The wife insisted that they keep this catch, despite the threats of the mermaid. Soon the sea swept in and carried off everyone except the once-wealthy widow.[12]

A lot of revealing information about merbeings in Eastern Europe was assembled by Finnish religious scholar Uno Holmberg in *The Mythology of All Races*. In the volume on Finno-Ugric mythology, he related how knowledge and stories of "water spirits" were widespread. Such treatments assume that these are only tales shared from person to person. The folklore collectors have marveled at the stories. Still, they have never been able to take the accounts about them of ordinary people seriously. The method is apparent in this paragraph about the Lapps:

> A god, known only at the coasts, is Akkruva, the upper part of whose body the Lapps imagine to be human, the head covered with long hair, the lower part of the body that of a fish. She rises at times from the sea and, sitting upon the water, rinses and combs her hair. Sometimes Akkruva walks up to the mouths of the rivers taking fishes with her, and at such times the catch is excellent. What this sea-spirit, called by Friis Avfruwa, really is, is shown above all by her name—a distortion of the "Havfru" of the Scandinavians, which, like the above mentioned being, had a human upper body whilst the lower body was fishlike.[13]

Holmberg further notes a wealth of traditions in Eastern Europe. A Lapp water spirit, to whom offerings are sometimes made, is the *Cacce-olmai* ("the Water Man"). He is the god of fishing, who brings fish to the hooks or in the nets and lines. In the notes of Simon Kildal, a missionary fluent in Lapp, we learn that Pagan roots survived attempts at conversion. Men made an image of the Water Man and put it into a crevice so that he might give them more luck. They also made sacrifices to the "Water Man" so that he should not do them any harm on the water.[14]

The *Cacce-olmai* of the Scandinavian Lapps corresponds to the *Cacce-jielle* ("Water dweller") of the Russian Lapps. This dangerous spirit calls upon and then tries to drag people into the water. The sight of it predicts disaster. A woman who saw this spirit while fetching water from the sea asked him whether his appearance foretold good or bad news. She was told that her son would die, which happened within three days.[15]

The "Water dweller" of the Russian Lapps corresponds to the *Vodyanoy* of the Russians, whose Water-Nymph, the *Rusalka*, is called by the Kola-Lapps *Cacce-jenne* ("Water mother"). She emerges from the water at dawn in the shape of a naked woman to comb her long black hair. When frightened, she throws herself into the water so quickly that she leaves her comb on the shore where she is sitting. She loves men and entices them to her. Kildal suggests the *Saiva-neida* ("Sea Maid") of the Western Lapps is Scandinavian in origin.[16]

One of the best summaries of Northern Europe's experiences with merfolk was written by Scottish ethnographer James A. Teit and published in 1918 in the *Journal of American Folklore*. This was part of a wide-ranging discussion of the sea-oriented traditions of the people in the Shetland Islands in the northern Atlantic Ocean. Here are some of the most informative passages about merfolk who adopt a fish-like appearance:

> These beings were like people when in their homes; but when travelling through the sea, they became half man and half fish; their upper parts remaining man-like, while their nether parts became fish-like, or were enveloped in a fish-like covering. Without this fish-like covering, these people could not travel the seas; but, as it was of no use to them in other elements, they immediately discarded it upon arriving home, and also when they came ashore in the upper world.
>
> They were about the size of the smallest people among us, and very well proportioned. Of a mild disposition, they were much attached to one another. They are known to have been fond of music, singing, dancing, and story-telling, and some of them played flutes and harp-like instruments.
>
> The men were darker in complexion than the women, and had hair and beards of various colors; for instance, brown, black, gray, and reddish. Their beards and hair were generally rather long. The women had fine features, light skins, and very long yellow hair, which floated around them when they were in the water. They sometimes came ashore in fine moonlight nights and sat on the rocks, combing their hair. They could sing very sweetly, and their singing enchanted men, and perhaps other beings. If a man

heard her song and saw her, he became spell-bound. It is said that men became so insanely in love with mermaids, that they followed them into the sea, and were drowned. When a mermaid sang, seals also came crowding around, and remained listening as if spell-bound. Mermen, unlike mermaids, very rarely came ashore.[17]

A 1739 report from Vigo, on Spain's northwest coast, suggests the merfolk could make their protective gear more hydrodynamic. Fishermen found a merman in their nets. He was five and a half feet tall and bald except for a beard. Some accounts refer to the merman's head as tapered. His back was hairy, with a very long neck. His arms were short and his hands longer and bigger than expected in proportion to the rest of the body. The fingers were elongated with claws. His toes were also elongated and webbed. His heels had fins "resembling the winged feet with which the painters represent Mercury."[18] He also had a fin at the lower end of his back, twelve inches long and sixteen inches wide. If the tapered head description is correct, the merman would form a conical teardrop shape, the most efficient hydrodynamic shape. With a dorsal fin for stability, and the feet acting as pectoral fins for maneuverability, this water ape would be fast and agile in the water. This suggests that *Homo sapiens* were not the only primates learning how to use tools.

Their fish-like skins were vital but removable articles of clothing.

> There seems to be some confusion in Shetland folk-lore between these sea-people, or *mar*-folk, and the *selki*-folk; as some people say of the former that they could assume seal-form as well as fish-form when travelling in the sea, or that they could more frequently assume the shape of a seal than that of a fish. In both cases real transformations were not involved, but mere coverings were adjusted to enable them to roam the sea. To travel underwater they enveloped themselves entirely in these contrivances; but on the surface of the sea their heads, necks, shoulders, and breast were uncovered, being out of water, and only the lower parts of necessity retained their fish or seal envelope. When they came ashore, they entirely discarded them, but never went very far; and in the case of alarm or some one approaching, they at once resumed their sea forms and

jumped into the sea. The loss of their possessions meant that they could no longer travel in the sea.[19]

The lore about the Selki appears to describe the activities of merbeings who have adopted the use of seal skins as their outfit of choice when living and traveling in the water world. The material about Finn folk appears to be explicable as the activities of human beings regarded as wizards. These particular people were not merfolk, but they did have a special relationship with seals. The difference between using fish skin and mammal skins (seals) might have created a significant cultural divide among merbeings. The distinctions in lore might reflect these differences.

ENDNOTES

1 Carrington, *Mermaids and Mastodons* (1957), 5.
2 Hudson, "A second Voyage or Employment of Master Henry Hudson, for finding a passage to the East Indies by the North-east." *Collections of the New-York Historical Society* v.1 (1811), 86–87
3 Weddell, *A Voyage towards the South Pole*, (1827), 142–43.
4 Pliny, *The History of the World* (1847), Book IX, Chapter V, p. 236.
5 Coggeshall, *Chronicon Anglicanum* (1875), 117–118.
6 Scribner, *Merpeople: A Human History*, 56.
7 "A Live Mermaid. And No Mistake," *Londonderry Standard*, 05 September 1838.
8 Leather, *Folk-Lore of Herefordshire* (1912), 168.
9 Simpson and Roud, *A Dictionary of English Folklore* (2000), 234.
10 Almqvist, "Of Mermaids and Marriages," *Béaloideas* 58 (1990), 4.
11 "Mermaid" *Galway Advertiser*, 25 September 1819.
12 Morvan, *Legends of the Sea*, (1980), 115.
13 Holmberg, "Finno-Ugric, Siberian" in *Mythology of All Races* v.4 (1927), 191.
14 Kidal, "Efterretning om Finners og Lappers hedenske religion," *Det skandinaviske Litteraturselskabs Skriftert* III (2) 1807, 458.
15 Holmberg, 192.
16 Kidal, 458.
17 Teit, "Water-Beings in Shetlandic Folk-Lore," *The Journal of American Folklore*, 31(120), April–June 1918, 189.
18 [Vigo merman]. *Universal Spectator and Weekly Journal*, May 5, 1739
19 Teit, 190.

Chapter Seven
Asia

THE MANY ETHNIC GROUPS OF ASIA have as many different names for merbeings, including *Samoyeds* (Master of the Water), *Ostiaks* (Water Spirit), *Voguls* (Water Khan), *Votiaks* (Water Man and Water Master), *Siryans* (Water Spirit and Water Dweller), *Cheremiss* (Water Master), and *Mordvins* (Water Dweller or He Who Inhabits the Water).

Holberg observes that both males and females are seen but avoid prolonged appearances.

> Like the male Water-spirits, the female has also features which betray her foreign origin. She is beautiful and her naked body is glistening white. Sometimes in the twilight the wife or daughter of the "Water man" will emerge on the shore to comb her long black hair. In some places she is said to have breasts as big as buckets. The male spirit, like the female one, is a shy being who immediately throws himself into the water on being observed by a human eye.[1]

Although predominantly associated with oceans in Europe and North American folklore, a mermaid can also be a freshwater spirit in Asia and Africa. Fresh water is obviously more important to humans, but could marine merbeings return to land to visit fresh water and be mistaken as a separate species?

The nomadic sea people known as the Moken inhabit the coast and islands in the Andaman Sea on the west coast of Thailand. A field study of the Moken adaptations to life on the water made an observation. After sustained exposure to saltwater, their skin developed a white, scaly layer that resembles the disease ichthyosis. The skin condition cleared when they returned to dry land and washed in fresh water.[2] This again begs the question: are ocean-dwelling merbeings similarly returning to the land to visit what African anthropologist Penny Bernard refers to as "living waters," the free-flowing fresh water where they are considered guardians and water spirits?[3]

People around the Caspian Sea were excited in 2005 about seeing a merman. He was seen in Iran and the Republic of Azerbaijan. The Iranians called him *Runan-shah*, or "Master of the Sea and Rivers." He was consistently described as being over five feet in height. He appeared to be a strong individual with hair both black and green. Other features noted were his large eyes, lack of chin, and four webbed fingers on each hand. He was seen by fishermen several times, both in the sea and near the shore.[4]

Theaters of combat often provide the opportunity for people to come into contact with unusual creatures. We find examples of these encounters in the Russian and Chinese military annals in Asia. Armed forces venture into rarely visited areas and bump up against the native inhabitants, who are often living fossils. For example, writer and aeronautics expert Martin Caidin describes an episode that took place in Vietnam in his book *Natural or Supernatural?*

The human participants in this encounter were utterly baffled by what they saw and what happened to them and were reluctant to talk about it afterward. In this case, this is probably a rare encounter with a Lizardman. Caidin was able to persuade a Marine to tell the story. By the time he spoke with Martin, David E. Gower was the sole survivor among the Marines who had experienced the encounter.

Gower related what happened on a mission with the Fifth Special Forces Group Marine Detachment. They went by boat north of the DMZ on December 17, 1974. They were dropped off and went one mile into the jungle to set up an ambush. It was a moonlit night, but in the jungle, it was dark. There they encountered some tall creatures that appeared to be glowing yellow. Gower saw three of them, ranging from seven to eight feet tall. He saw they had three fingers on each of their hands. He saw long claws on the fingers. And their tracks, which they could examine, showed three toes.

Gower and his fellow Marines pieced together a description of the head of the Lizardmen as having "two eyes, some sort of nose slit on flat faces, and audio holes on the upper part of the sides of their heads."

Gower decided to withdraw his six-man detachment from the area. At that point, the "Big Yellows," as he called them, began running directly after the men. Gower fired upon one of them with little effect. The Marines retreated to their boat and left the area.[5]

If the "Big Yellows" were glowing blue instead, one could argue that the Lizardmen had passed through bioluminescent algae, common in Vietnamese waters. Yellow is a different matter. For a water ape to glow yellow, it must be deliberate. Although some seaworm and octopus species can produce yellow bioluminescence, Lizardmen are swamp dwellers. It seems more likely the "Big Yellows" have weaponized the ample supply of fireflies and their larva, crushing them and painting themselves with the bioluminescent viscera. The effect would be short-lived but effective, as Gower and his marines could attest.

There are other theories. Biologist and cryptozoologist Ivan T. Sanderson described several cases of fishermen with tumors infected with bioluminescent sea animalcules.[6] The luminescence suggests that this type of merfolk could be defensive or aesthetic. If not the specific color, the concept has been utilized on both sides of the Pacific. For example, a 1972 sighting at Thetis Lake in British Columbia mentions the creature being "fluorescent silver-gray" in appearance.[7]

In May of 1846, a local government official went to investigate a mysterious green glow in the sea on the Japanese island of Kyūshū. He discovered Amabié, a similarly bioluminescent creature, emerging from the water. She had fish scales, a beak, long hair, and three fin-like legs. Green bioluminescence would suggest the Japanese lantern fish, abundant in the waters off Kyūshū.

Amabié is considered *yogenjū*, a prophetic beast hybrid. She predicted the future epidemic in 1846 and advised that it could be prevented by the widespread distribution of her image. An earlier example is *jinja-hime*, with a fish body and a woman's horned head. She is also said to have appeared in 1819 to predict an epidemic.[8] Despite having three legs, Amabié has steadily evolved into a mermaid since the 1980s. When the Covid pandemic struck Japan, the mermaid version of Amabié became the reassuring image on government posters with helpful preventative tips.[9]

The Japanese equivalent of a mermaid is the *ningyo*. *Ningyo* are usually described as unattractive. The descriptions usually include razor-sharp teeth and, occasionally, horns. Of course, like Amabié, the *ningyo* have become progressively more similar to the stereotypical Western mermaid.[10] The earliest reference in the *Nihon shoki*, Japan's second oldest surviving history. The *Nihon shoki* reports that in the summer of 619 AD, a child-sized creature was caught in a fishing net in the Gamō River in present-day Shiga Prefecture. The creature was neither a fish nor a man.[11]

The remains of a *ningyo* can be seen at the Ryuguji Temple in Fukuoka's Hakata Ward. A ningyo is said to have washed ashore at Hakata Bay on April 14, 1222. The local shaman considered it a good omen, and the mermaid was buried in Ukimido. Only six pieces of the bones, which are said to have been exhumed between 1772 and 1781, survive.[12]

The mummified remains of a second *ningyo* are venerated at Enjuin temple in Asakuchi, Okayama Prefecture. The foot-long mermaid was caught in a fishing net off Kochi Prefecture between 1736 and 1741. It passed through several owners before it was given to the temple. It was displayed in a glass case until the 1980s, when it was moved into a safe for climate control. The mummy still has nails, teeth, hair on its head, and scales on the torso. It was first documented in the 1920s by Kiyoaki Sato, a natural historian and local folklorist. Sato wrote Japan's first encyclopedia on *youkai*, a grouping of Japanese folklore that includes both supernatural creatures and those that today would be considered cryptids.[13]

This mummified mermaid unexpectedly returned to the news when the temple agreed to allow noninvasive testing of the body.[14] The journalist covering the mermaid's recent CT scan also notes mermaid mummies are also venerated at Mount Koyasan in Wakayama Prefecture and on Amami-Oshima Island, Kagoshima Prefecture.

In February 2022, the mummified remains were brought to Kurashiki University of Science and Technology. The mummy was examined under high magnification and given CT scans and x-rays. At the same time, a team of historians delved into mermaid lore. A year later, the results were released. It was an example of what in the United States is known as a "Feejee mermaid," not actual mermaid remains. The upper half has no skeleton. Bones were found primarily in the head: a carnivorous fish's jaw and teeth. The skin is from a puffer fish coated with glue made of charcoal and sand. The lower body contained bones from a tail or dorsal fin, covered with skin from a croaker (any fish of the family *Sciaenidae*). Radiocarbon dating on the mummy's scales determined it was likely made in the late 1800s. The mermaid's origin relies on a hand-written history from the 1870s, so the mummy was created at the height of the Feejee mermaid trade.[15]

Historically, the final report confirmed twelve additional mermaid mummies, with two additional relics confirmed in Okayama Prefecture during the project. The one at Enjuin temple was the first that has been closely examined. The report notes these artifacts were being made since the Edo period (1603–1868). They were good luck talismans meant to represent the elusive *ningyo*, a desperate attempt to cure diseases like smallpox and measles introduced by occidental explorers.

Kozen Kuida, the chief priest at Enjuin temple, was philosophical. He said the mermaid mummy would remain a prized possession of the temple. "Many people in this area came here and joined hands to pray (to the mermaid), so it holds their thoughts," Kuida said at a news conference.[16]

Another report came from crew members aiming to salvage the steamship *Takachiho Maru*, which went aground on the rocks off Dzususaki, Tsushima, in heavy fog on May 11, 1891. Salvage crews tried for six weeks to secure the vessel and raise her. By mid-June, the *Takachiho Maru* was considered salvageable, thanks to unusually fair weather.

On September 10, one of the workers described an encounter where he killed a mysterious creature.

> [The worker] observed a very singular looking fish and immediately obtained a spear, with which he killed it. On *examining he found that it had a girl's head,* with human eyes, mouth and ears. Hair was growing on the head and there were appendages like hands where the pectoral fins should have been. All this part of the thing was undoubtedly human. The belly, back and caudal portion were piscine, their color a brownish black, like the color of the carp. The monster was two feet long, and measured 1 foot 9.5 inches in circumference. Here was one of the creatures known as a mermaid (ningyo). Tradition says that anyone eating a mermaid shall live 1000 years. The fisherman accordingly cooked and ate it, finding it exceedingly palatable, much superior in taste to a bream or a carp.[17]

The evening after the *ningyo* was killed and consumed, a gale hit Tsushima. This confirmed to the salvage crew that their co-worker had indeed killed a mermaid. The *Takachiho Maru* was damaged in the storm beyond repair and was declared unsalvageable.

The size of the *Takachiho Maru* creature seems small for *ningyo*, let alone a water ape/mermaid. Is there another cryptid involved, a fish with a human-like head? There are similar reports from Louisiana.

That mermaids inhabit the Gulf of Mexico in the opinion of Tulane University scientists and authorities at the Louisiana Historical Society, after the examination of a skeleton turned over to the society to-day by Captain C. A. Thompson, United States Government lighthouse keeper in Pass-a-l'Outre, near the mouth of the Mississippi River.

The skeleton includes the windpipe and spinal column, both of the same structure and in the relative position as might be found in a human. The bones of the breast and of the pelvis are similar to those of a woman. There is no leg formation, the lower extremity having been apparently provided with a tail. A small hinge bone became detached, and bone experts say this once suspended a tail. The trunk is five inches long and the head is of graceful outline.

Captain Thompson, who found the woman-fish lifeless on a small island in the Gulf off the river's mouth, boiled the skeleton to separate the flesh from the bones. In the company with Caption William A. Meyers, of New Orleans, master of the packet *El Rito*, he came to the city this afternoon with his curious discovery.[18]

Another encounter with a fish/human hybrid occurred in California's San Francisco Bay Area.

Deputy meat and milk inspector Downie confiscated a strange fish in the free market last week as unfit for food. It has the usual tail, but has two short four legs and a head almost human in shape, the nose, mouth and chin being clearly defined. It is about 18 inches in length and thick as a man's forearm. Downie is certain that he has in his possession a specimen of the fabled merman. The fish was caught in the Oakland estuary.[19]

This mysterious "mermaid fish" requires further examination. It is possible that the manufactured fakes of the "Feejee Mermaid" (made popular by the American showman and famed hoaxer P. T. Barnum) were based on this cryptid mermaid fish and not *ningyo* scaled down in size.

In 1560, the Jesuit mission in Goa, India, sent a group of priests led by Father Henrico Henricio to Manar Island off Ceylon (Sri Lanka). They were there to establish a hospital in the aftermath of a war with the Jafana-patanian kingdom of Ceylon. Dimas Bosque, the physician to the viceroy of Goa, was also sent.[20]

As they tended to the sick, a fishing crew arrived, begging Father Henry to go to their boats. Henricio and Bosque agreed. In astonishment, they found sixteen mermaids trapped in the *cheena vala*, a stationary lift net submerged to a certain depth and lifted vertically out of the water. Nine females and seven males were carefully examined, and the survivors were set free. The ones that died were dissected.

The head was a round shape, with a short neck. The ears were identical to humans. The eyes were also human in location and color, which surprised the witnesses, expecting a more piscine appearance. The nose and lips were very similar in size and appearance to humans. The teeth were not serrated as expected but were even and unusually white. The mermaid breasts were not pendulous like mature women but more spherical like a youth. The mermaid bodies were longer than that of the mermen.

Bosque did note the arms were not round but widened as if they were made for swimming and were three feet long, slightly longer than humans. The hair was very soft and long on both genders. Genitalia, both external and internal, showed no noticeable difference from humans. Since the tail is not mentioned, it can be assumed these merbeings had legs.

The Dutch artist Samuel Fallours served as a minister's assistant from 1706 to 1712 in Ambon, Indonesia. A skilled artist, he would illustrate the local fauna to send back to the Netherlands. One illustration he drew was of his mermaid. According to his notes, his son purchased the "syrene" from islanders on Boru in the neighboring Maluku Islands.[21] The siren survived four days in a tub of water at Fallours' home in Ambon before it starved to death. He drew his illustration from the remains and involuntarily "donated" the corpse to Governor Van der Stel, who had it shipped back to the Netherlands as a royal gift. Fallours sent his picture of local fish (and the siren) to Amsterdam. The mermaid was described as follows.

> [The mermaid had] a head like that of a female human being, having balanced, well-proportioned parts, eyes, nose, and mouth; however, the eyes, being very light blue, appeared to be a little different, like those of another kind of human being. The hair, which extended down over the neck, was sea-green and gray. She had breasts, long arms,

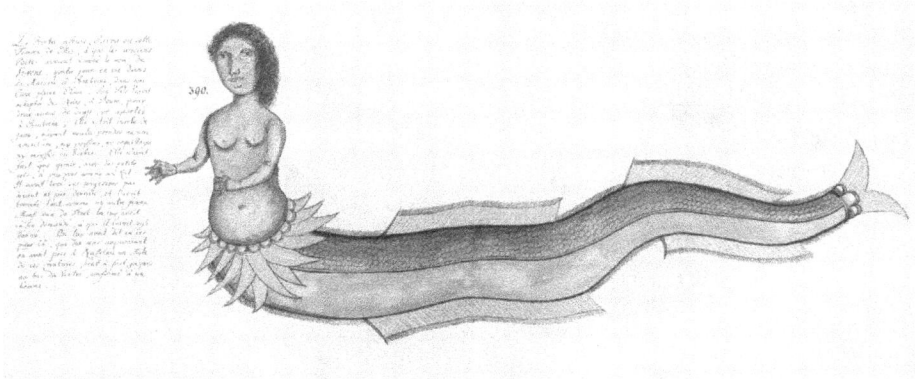

Samuel Fallours's painting of the Indonesian "Sirenne," in *Poissons, Ecrevisses et Crabes* by Louis Reynard, 1754.

hands, and the other parts of the upper-body, just like, and almost as pale in color as, any other woman . . . but the body below the navel looked like the lower part of a fish . . . with scales like those of a carp.[22]

A thought-provoking detail in the drawing is that each wrist has a clearly drawn line around it as if the webbed hands were some sort of gloves.

Fallours was an assistant to the minister and his immediate superior in Ambon was Dutch missionary and naturalist François Valentijn. In 1714, Valentijn was returning to Europe from Batavia (now Jakarta, Indonesia), a grueling fifteen-thousand-mile sea trip. On the morning of May 1, 1714, off the west coast of Africa near modern-day Liberia, Valentijn also saw a mermaid. He later wrote that the crew first thought it was a shipwrecked sailor treading water in a gray sweater and a woolen Monmouth cap.[23] As the ship came within four hundred feet of the creature, it was apparent it was not a human as the mermaid dove beneath the waves. Fallours' "syrene" is included in Valentijn's five-volume history of the Dutch Indies, as is his illustration.

Valentijn admitted he had not seen Fallours' mermaid, but it was not the only one he had heard of in the islands. He repeated a story told him by a Dutch East India Company lieutenant. The lieutenant was at a village in the Nau Binau region of Ambon when he saw two merbeings. They were swimming in the bay, side by side, which he believed indicated they were male and female. Over fifty witnesses saw the two aquatic creatures again six weeks later. He described the color of the mysterious beasts as greenish-gray, similar to the accounts by Fallours and Valentijn. The lieutenant added they appeared human to the waist, with arms and hands, but the torso terminated in a point. One was larger than the other, and they had long hair.

We have some valuable testimony from the Pacific Ocean, courtesy of someone who saw water apes on three separate occasions. His knowledge of water apes was based on all those experiences. William Marks, author of *I saw Ogopogo!*, interviewed the witness, Rein Mellaart of Penticton, British Columbia. During World War II, Mellaart was stationed on Morotai Island in Indonesia when he had an encounter.

> One day I heard an awful commotion and yelling from some of the native fishermen coming in, dragging a net. It looked to me at first as they had caught a shark. The thing made an awful splashing and thrashing around in a vain effort to escape. But as they came closer, it seemed to me that they had a human being in their net. Then as I watched, one minute it seemed human and the next it appeared to be about a seven ft. long shark. 'What is it,' I cried. One of the natives answered in broken English "We catch mermaid again."[24]

Mellaart told them to let it go, but they refused. They said they would not kill it but let it die on the beach. The creature tried for half an hour to escape and then quit. It then began to cry. Mellaart went for help to intercede, but the mermaid had died when he returned.

> The natives had caught some of her tears and superstitiously believed it would work as a love potion. The bottom part of this strange creature was scaly, just like a fish. But from the navel up, it looked as human as any person you'd meet on the street. Contrary to reports, however, it was not the beautiful siren that sailors talk about. The features were coarse, and she had a long pointed nose. But the complexion was the most beautiful you could imagine, a lovely pinky red.
>
> The fish-woman had lovely hair, just about even with the fish part. It was thick and when they swam on the surface, it streamed behind them.

These mermen and mermaids travel in schools (not too many) as a rule. They are terribly frightened if a person or boat comes near them. Natives that did manage to get near them, claimed they signalled each other and dived to a great depth. They'd watch and maybe see them come up a half mile or so away for air, then dive again.

The bottom part of the mers that I saw was exactly like a dolphin, with a double fin on the end. Another peculiar thing about them was that each one had 6 fingers, or I would say four fingers and two thumbs on each hand. With these hands, they drag themselves up on the beach, generally at night.[24]

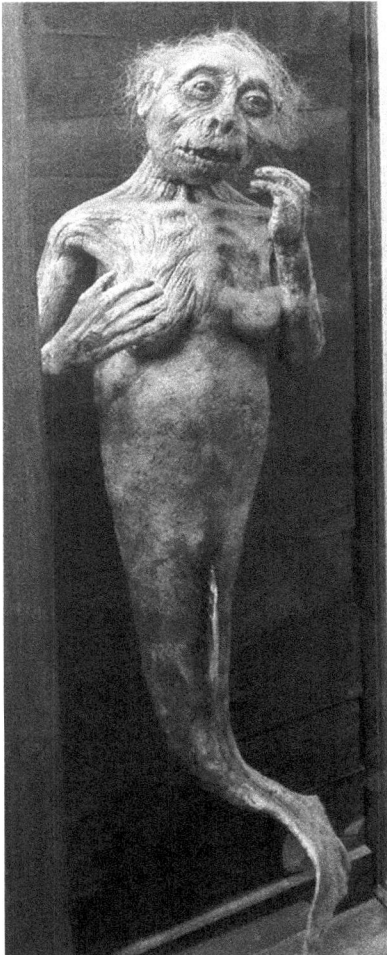

Shan hai jing (Classic of the Mountains and Seas) is China's earliest cultural and geographical record, written over twenty-five centuries ago. It introduces the *lingyu* (man fish), an ocean-dwelling creature with a human face and a fish's body that lives in the sea. The book further notes other types of mermaids, such as the *renyu, chiru, diren*, and *huren*. The book doesn't specify physical differences, so the variety may indicate different groups of water apes or other names regionally among the Chinese. In either case, it suggests a wide area of coastal inhabitation. It does not, in either case, indicate the friendliest of relationships. According to Sima Qian's *Historical Records*, candles in the tomb of the First Emperor Qin were made from the fat of

On July 17, 1842, the news broke that P.T. Barnum had acquired the mysterious FeeJee Mermaid; it was later destroyed in a fire. The pictured FeeJee Mermaid replica (created by Erik Gosselin of LifeMakerFX, Quebec), from the film *P.T. Barnum* (1999), is on exhibit at the International Cryptozoology Museum in Maine.

the *lingyu*.[25] The candles were supposed to burn for a long time, appropriate for an emperor's tomb. Still, it could not have improved relations with the water-dwellers.

Farther north, humans and mermaids seemed more amiable. Some of the earliest explorations by Jesuit missionaries made first-hand observations of a culture along the Amur River above the mouth of the Dondon River known as Yupi Tartars (Fish-Skin Tartars) or the Hezhen people.[26] Although the ethnic group was decimated by Japan's occupation of Manchuria in the 1930s, they were legendary for their skills in fishing. So much so that a legend arose that the Huzhen were descendants of mermaids. Legends aside, the Fish-Skin Tartars got their name because they made garments from carp, pike, and salmon skin. One of the few surviving makers of fish skin clothes explained the process. The skins were dried and processed repeatedly through a wooden press for a month. Creating a shirt and pants requires over fifty fish, resulting in a leather outfit with a crisscross pattern that makes it more pliable and durable than most leathers.[27] Perhaps the term "descended from mermaids" was originally "a close relationship with the local mermaids," garbled by time. It could be that the Huzhen were asked to help manufacture protective gear for the apes using techniques developed by the water apes. This could explain the fish-like descriptions. A viable exchange network among the sea apes could create sufficient demand for fish-skin goods; it would be prudent to involve the Huzhen in a mutually beneficial business relationship between humans and apes.

In 1901, British missionary John Batchelor published his second condescending look at the Ainu people of northern Japan around Hokkaido. Filtered through the missionary's biased perspective, everything in Ainu culture was dualistically good and evil. He mentions the mermaid, which can change form. Among the forms the mermaids can assume is one that sounds like a water ape removing its protective "mermaid gear." Mermaids, according to the Ainu, can take human form.[28] The Ainu also settled on the Russian island of Sakhalin, thirty miles across the Sea of Okhotsk from Hokkaido. The mouth of the Amur River is twenty-five miles across from the northern part of Sakhalin, meaning the "transforming mermaids" of the Ainu and the Fish-Skin Tartars were in contact.

During the Qing Dynasty, a folk artist, Nie Huang, traveled along the coasts, illustrating over three hundred marine creatures, including a merman, in his four-volume *Hai Cuo Tu*. With black skin and yellow hair, the merbeing in his book has webbed hands, a short tail, and a human head. Although usually translated to say the merman has red "wings" on his back, the illustration shows an extensive red dorsal fin.[29] This modification could be the protective suit of a water ape, added for stability.

In 1739, Cantonese writer Fan Duanang jotted down his observations of the Cantonese rivers and the South China Sea.[30] He includes two stories about mermaids. First, a fisherman captured a mermaid off the coast of what is now Hong Kong's Lantau Island. She was human in appearance except for being covered in fine hair of multiple colors. The author further notes that the creatures were frequently seen in the waters around Lantau Island and the South China Sea, especially near the Zhujiang River.

> The Cabinet Councilor Cha Tao being dispatched on a mission to Korea, and lying at anchor in his ship at a bay upon the coast, saw a woman stretched upon the beach, with her face upwards, her hair short and streaming loose, and with webbed feet and hands. He recognized this being as a mermaid (or man fish) and gave orders that she should be carried to the sea. This being done, the creature clasped her hands with an expression of loving gratitude and sank beneath the waters.[31]

These encounters appeared in an 1876 book by Nicholas Dennys, a British diplomat in Singapore. The two accounts are widely cited, but Dennys did not author those sections. This is relevant because the original article in 1869, which Dennys used as source material, was by W. F. Mayers and includes additional details of interest that Dennys inexplicably edited out.

Mayer's translation of the Lantau Island mermaid meeting differs. It specifically notes the lower part of her "body was covered with fine hairs, of many beautiful colors, one or two inches in length, which served as a covering to her nakedness." Mayer's original text also elaborates on the Korean mermaid. He describes her as four-to-five feet in length, "and possessing the human form, hair, and distinctions of sex, but having a short finny ridge upon the back, of a pale red colour, which marks them as belonging to the order of fishes."[32]

In both cases, we have human-like bipedal creatures showing adaptations for aquatic life. The colorful hair description could indicate an aesthetic consideration or may be related to the luminosity discussed earlier. Regardless, it also provides another account of a red dorsal fin/ridge modification from an account a century after the report in *Hai Cuo Tu*.

In 2009, Japan-based researcher Brent Swancer came across an un-known account from WW2. This encounter was so obscure that no one remembers it, even in Japan, and it relates to what's called the *"Orang Ikan"* (オランイカン).[33]

In 1943, the Japanese occupied the Kei Islands, part of the Maluka Islands of Indonesia. Members of the military observation team reported seeing something vaguely human in the ocean. A composite of accounts suggested a creature with a face similar to a human. It was about five feet long with pink skin and spines on the head. It had arms and legs with no tail.

The local villagers were familiar with the creature, telling the team that they were known as the *Orang Ikan*, or translated from Malay into English, "Fish People." The villagers ignored them except when they got caught in the fishing nets.

The Japanese were informed the next time a corpse of an *Orang Ikan* had washed ashore, and a Sergeant went to the village to examine the re-mains. He described it as having a headful of red-brown, shoulder-length hair and spines along the neck. The face was said to resemble a human or an ape with a short nose, broad forehead, and small ears. The mouth had no lips and was specifically compared to a carp. It was wide and filled with tiny, needle-like teeth. The creature's fingers and toes were long and webbed.

Swancer discusses the perceived similarities in the description to the Thetis Lake Monster and the Loveland Frogman. He suggests that there could be a marine-adapted population of *Homo floresiensis*. It is an inter-esting hypothesis, though it doesn't negate the possibility of *Oreopithecus* being the most likely candidate for the ancestor of the water ape.

ENDNOTES

1 Holmberg, "Finno-Ugric, Siberian" in *Mythology of All Races* v.4 (1927), 195.
2 Gislén and Schagatay, "Superior Underwater Vision Shows Unexpected Adaptability of the Human Eye," in *Was Man More Aquatic in the Past?* (2018), 170.
3 Bernard, "Mermaids, Snakes and the Spirits of the Water in Southern Africa" as quoted in Varner, *Creatures in the Mist* (2007), 16.
4 Garifdjanov, "Mysterious amphibious human-like creature spotted in the Caspian Sea." *Pravda (Moscow Russia)*, 25 March 2005.
5 Caidin, *Natural or Supernatural?* (1993), 126–143.
6 Sanderson, "Luminous People and Others," *Pursuit*, July 1973, 66–67.
7 "'Monster' Seen in Lake," *Daily Colonist* (Victoria, BC), 22 August 1972.
8 Nagano, "Amabiko the prophetic beast, revisited" in *The Forefront of the Study of Yōkai Culture* (2009), 131–162.
9 Furukawa and Kanssaku, "Amabié—A Japanese Symbol of the COVID-19 Pandemic," *JAMA*, 11 August 2020.
10 Hayward, "The Mermaidisation of the Ningyo" in *Scaled for Success* (2018), 51–68.
11 "Nihongi: Chronicles of Japan from the Earliest Times to A.D. 697," *Transactions and Proceedings of the Japan Society* 1, Supplement 1 (1896), 147.
12 Sadamatsu, "'Mermaid bones' from 13th century keep legend alive in Fukuoka," *The Asahi Shimbun* (Osaka, Japan), 13 February 2017.
13 Kiyoaki Sato, "Genkou Zenkoku Youkai Jiten" *Hogen Sosho* 7, (1935).
14 Kunio Ozawa, "Scientists try to unravel mystery of eerie 'mermaid mummy.'" *The Asahi Shimbun* (Osaka, Japan), February 19, 2022.
15 "圓珠院所蔵「人魚のミイ」研究最終報告 (Final report on research on the "mermaid mummy" in the Enju-in collection)," February 7, 2023.
16 Ozawa, "Tall tale: Study finds 'mermaid mummy' largely a molded object," *The Asahi Shimbun*, February 8, 2023.
17 "Killed a Mermaid," *Galveston Daily News*, 28 Sept. 1891.
18 "Mermaid Found at Last." *New York Tribune*, 10 July 1911.
19 "Mermaid Found in San Francisco Bay," *Sausalito News*, 29 June 1907.
20 Sacchini, *Historiæ Societatis Iesu pars secunda siue Lainius* (1620),Volume 2, part 4, 162 (no. 276).
21 Pietsch, "Samuel Fallours and his 'Sirenne' from the Province of Ambon" *Archives of Natural History* (February 1991), 1–25.
22 Valentijn as translated in Pietsch,7–8.
23 Valentijn, *Beschryving van Oud en Nieuw Oost-Indië* (1724), 331–332.
24 Marks, *I Saw Ogopogo!* (1971), 18–20.
25 Ssu-ma Ch'ien, *The Grand Scribe's Records* (1994), 155.
26 Du Halde, *Description Géographique, Historique, Chronologique, Politique et Physique de l'empire de la Chine et de la Tartarie Chinoise*. Vol. IV (1763), 43.
27 "China 'Mermaid descendants' weave garments from fish skin," *The Asahi Shimbun*, January 21, 2020.
28 Batchelor, *The Ainu and Their Folk-Lore* (1901), 541.
29 Zhang Chenliang. *Chinese National Geography - Notes of Hai Cuo Tu* (2016).
30 Fan Duanang. *Yue Zzhong Jian Wen* [1801].
31 Dennys. *The Folk-Lore of China* (1876), 114–115.
32 Mayers, "Mermaids and Mermen in the Chinese Seas" *Notes and Queries on China and Japan* (July 1869), 100.
33 Swancer, "Orang Ikan," *CryptoZooNews*, July 17, 2009.

Chapter Eight
Africa & the Middle East

SOUTH AFRICAN ANTHROPOLOGIST Penny Bernard has found that among many of the Southern African Indigenous people, water is essential to both spiritual and physical life.[1] Spirits manifest as snakes and mermaids (often interchangeably) who reside in the water, interacting with humans as needed. The rivers and oceans are the dwelling places of such interactions, making the water sacred to many African healing traditions.

> As a result of the profound sacred status that the many rivers, pools, and water sources hold for Southern African Indigenous communities, there existed in the past, and to some extent today, a range of taboos surrounding their access and utilization. Pools, rivers, and expanses of water are held with a mixture of awe, fear, and reverence. Great care was taken in the past to avoid disturbing or angering the water spirits. Common people were forbidden to go near sacred pools where the snake, mermaids, and spirits were known to exist. This injunction was reinforced with the fear that uninvited people would be taken under the water never to return.[2]

The pan-African water goddess *Mami Wata* epitomizes this interchangeability between mermaid and serpent and the evolution of local water spirits into a singular figure. She may appear as a beautiful woman, a mermaid, or a snake from the hips (an African equivalent of the South Asian naga, a half-human, half-serpent creature). Whether this indicates a co-mingling of local versions of water spirits or that the water apes could weaponize serpents to prevent humans from disturbing them remains to be determined. Art Historian John Henry Drewal traces the introduction of the European image of mermaids to West Africa back to figureheads on Dutch and other European ships in the pre-colonial era.[3] Because early European travelers were associated with the sea spirits from the 15th century onwards, a new water spirit arose as *Mami Wata*. Drewal also believes the most common representation of *Mami Wata* today solidified in the late 19th century after a tour by a German circus widely distributed a poster of a snake charmer.[4] A critical detail to remember with Drewal's work is that he is looking at the evolving image, not the story's original source that so quickly became *Mami Wata*. Drewal notes finding versions of *Mami Wata* in twenty African countries and fifty cultures.[5]

The Bishop of Bergen, Erik Pontoppidan, is best known for introducing the Kraken to the world in his 1752 book *The Natural History of Norway*. But Pontoppidan also discussed mermaids. He was deeply concerned about the species. He said a soulless animal should not appear so similar to a human because man was created in God's image. He finally decided that a merman should be called a sea ape to downplay similarities to mankind. If only he knew how close he had come to the truth so many centuries ago.[6] Since apes were animals, not men, their mistreatment is more palatable to European theological sensibilities. Pontoppidan then quotes a 1668 passage on sea apes as a delicacy in Angola, quoting Dutch geographer Olfert Dapper.[7] Fortunately for the merbeings, the story is incorrect. Dapper uses the term "*meerminne*," which is usually, but not exclusively, used as a literal translation of mermaid. Pontoppodan missed Dapper's elaboration on how the Portuguese referred to the creature as "*Pezze Mouller*," a garbled phonetic rendering of "fish woman" in Portuguese. This term explicitly meant West African manatees. So, this is one of the few times that claiming a mermaid is a manatee is correct. It is good news for merfolk but bad news for manatees, which Dapper says taste like pork.

In Nigeria, Benin folklore has mermaids as handmaidens of the river goddess *Igbaghon*. The mermaids similarly take interlopers, such as those coming to fetch water. They are never seen again on the surface.[8] This concept of uninvited or unworthy visitors being taken below the surface is distinctly similar to that of Native American stories. Eurocentric mermaids are more random in selecting victims to drown.

The Shona of Zimbabwe identify mermaids, known as *njuzu,* as spirits of nonhuman origin. This alienness is considered a fundamental part of the *njuzu*.[9] Similarly, the Cameroon water spirits, the *mengú*, are, as missionary Johannes Ittmann puts it, "human-like yet not human, fish-like yet not fish; they are smaller than humans, can walk like them, and swim like fish." Ittmann may be conflating coastal dwelling *mengú* with pygmies. Ittmann also notes the *mengú* are unattractive. They are covered in hair with oversized mouths that open almost ear to ear. The eyes are also disproportionally large. Yet, in other cases, they are considered supernaturally beautiful.[10] This difference in appearance could be due to sightings of primates seen in and out of their protective suits.

Zimbabwe mermaids continue to this day. In 2012, work on planned reservoir upgrades stopped because mermaids had been scaring workers away. Water Resources Minister Samuel Sipepa Nkomo notified a parliamentary committee that workers refused to work on Osborne Dam in Manicaland and Gokwe Dam in Midlands.[11] These were only two of many reports of mermaid incidents. In 2010, an overloaded speedboat capsized at Chiota Dam, drowning three passengers. Survivors said the boat was following a snake-like creature.[12] Tribal Chief Makope was not surprised as the dam was home to a mermaid, and the locals knew to avoid the area of the dam and considered it sacred. In all cases, the mermaids were appeased when the water ministry hired traditional healers and leaders to conduct rituals in which they slaughtered cattle and brewed beer.[13]

This collaborates anthropologist Penny Bernard's specification that the mermaids/snakes only inhabit "living waters," flowing waters such as rivers, the ocean, or deep pools near waterfalls, a concept that spans the continent.[14] At least in Zimbabwe, it also includes reservoirs.

In South Africa, every decade or so, the residents of Suurbraak report a mermaid in the Buffelsjags River. The most recent sighting was on January 5, 2008.[15] Suurbraak resident Daniel Cupido was relaxing along the river with friends and family. They heard what Cupido thought sounded like "bashing on a wall." Suspecting vandals, Cupido and his friend followed the noise to a local bridge where he saw a "white woman with long black hair thrashing about in the water." Fearing she was injured or drowning, Cupido waded toward her but said he reversed direction when he noticed a reddish shine in her eyes. Another member of the would-be rescue party said the woman "had an eerie silver-white glow," similar to that of the phosphorescent lizardmen of Vietnam and the Japan yogenjū discussed in Chapter Six.

A famous rock-painting scene in South Africa is the "Mermaid Frieze" at Ezeljagdspoort, on the Braek River near Oudtshoorn. The

Rock painting of mermaids (watermaidens) at Ezeljagdspoort, South Africa, from *Narrative of a Voyage of Observation among the Colonies of Western Africa* by James Edward Alexander (volume II, plate III), 1837.

image was drawn by a member of the Bushmen, the Indigenous peoples of Southern Africa, among the oldest cultures on Earth. The rock art was first published in 1837 and shows a semi-circle of over a dozen silhouettes of creatures with long arms and fishtails.[16] From the beginning, scholars debated whether the drawings were symbolic, metaphors, shamanistic visions from a "rain dance" trance, or something else.

It is not the only rock painting that suggests the Bushmen believed in water maidens or *watermeides*. Rock paintings that depict water people with legs and tails were seen in Upper Cave in Mangolong.[17] The fish-tailed men (*Qweqwete*) live underwater and are shown capturing a snake.

In 1970, J. Leeuwenburg of the South African Museum located a long-lost transcript of an interview with an elderly Bushman. The interview was conducted by an anthropologist in 1875, and discussed the *watermeides* who lived underwater and lured people to doom.[18] Leeuwenburg concludes that the Bushmen had an extensive collection of folktales and that pictorial representations of these tales can be found among the paintings. In other words, the "Mermaid Frieze" and the "Fish-tailed Men" rock paintings may simply be exempla, a visual reminder of a story. It doesn't determine whether the tale is based on a true story.

Today, Southern African mermaids have a more complicated relationship. The water spirit manifestation of the mermaid has become blurred with that of the *Ilomba*, a snake magically conjured. The snake feeds on blood and steals people's souls. The *Ilonba* is feared and hated by the Barotse people, particularly in the Western Province, Zambia's least-developed and isolated region. In January 2020, the Bethel Church in Chipata Overspill, Lusaka, was burned by an angry mob that was whipped into a frenzy by a local witch doctor who claimed there was a mermaid *(ilomba)* in the church.[19]

Even though the term mermaid is used to describe these creatures, these are not our water ape/mermaids, but rather, a new folklore hybrid creature.. African local journalists have also referred to *ilomba* as a goblin, further muddying the waters. Still, it is disconcerting to see modern newspapers where every drowning, illness, and missing person is suspected to be a victim of "mermaids."

This renewed but misguided interest in mermaids seems to date back to the 1970s when the South African press became interested in discussions about rock paintings. In a letter to the editor of the *South African Archaeological Bulletin*, a reader disagrees with an article reinterpreting the "Mermaid Frieze." The reader argues that not only did the Bushman believe in water people, but that a mermaid was spotted in a storm sewer at Labala in Lusaka, Zambia, in December of 1977. Witnesses said she looked like a "European woman from the waist up, whilst the rest of her body was shaped like the back end of a fish, and covered with scales."[20]

Mermaids appear in Arabic records before any written reports in Europe.[21] In modern Arabic, *khayilaan* is the half-human/half-fish, with other specific terms for specific regions, with differing interactions with land-dwellers, such as the carnivorous *aeisha qandesha* of Morocco.

Arab historian and explorer Abu al-Hasan al-Mas'udi includes accounts of sailors encountering the seductive *banat al-maa,* who are more traditional in appearance: fish bodies with human faces, breasts, and long hair. Sightings of *banat al-maa* extend from the Mediterranean to lakes close to the Nile Delta.[22]

Al-Mas'udi was not the only early Middle Eastern scholar to mention merbeings. In his book, *Ajā'ib al-makhlūqāt wa-gharā'ib al-mawjūdāt* (*Wonders of Creatures and Marvels of Creation*), Zakariya al-Qazwini, relates several tales of interest.

In 894AD, for instance, Prince al-Muktafī bi-llāh went fishing in the Caspian Sea. In the course of the day, a large fish was caught. The fish was cut open, and a mermaid was found in the fish's belly. She was still alive. She alternated between clutching her face and tearing out her hair.

She made a few sighs and died a few moments later. The only additional description was that she was wearing a pair of pantaloons made of what appeared to be human skin made with such craftsmanship there were no seams in the garment.[23] Where al-Qazwīnī describes a mermaid clutching her face, it could be that a water ape was suffocating in protective gear, desperately attempting to remove the headpiece before collapsing from asphyxiation. The seamless wardrobe speaks for itself in light of the Yupi Tartars of the Pacific Coast (see Chapter Seven).[24]

Al-Qazwini also relates how a contemporary found the mummified remains of a merman and put them on display. The man who found the creature said he came from the Mediterranean Sea. This obscure case has rarely been discussed in cryptozoological circles. As popular as Qazwīnī's book has been through the centuries, it has never been reprinted in English. It is through other authors that the reports came to light in the West.

Mermaids from the Mediterranean may have traveled upriver further than expected. In 1837, German bon vivant and travel author Prince Hermann von Pückler-Muskau visited antiquities along the Nile in Upper Egypt.[25] One evening, he discussed folklore with a Dongolese servant of the Pasha. The Dongolan assured the prince that beyond Old Dongola in what is now the Northern State of Sudan, there lived enchanted Albinos and secret cannibals. He also added that in Sennaar, there were still "syrens." He knew this because he had personally seen them more than once.

Sennaar is a town in southern Sudan on the Blue Nile, one of the major tributaries of the Nile as it flows into the Mediterranean. The Nile, which flows northward for four thousand, one hundred and sixty miles from east-central Africa, has a drainage basin that covers eleven countries. This explains how landlocked equatorial Burundi has a mermaid tradition. In isolation, the Burundian mermaid has evolved differently from the human-friendly (or at least human-tolerant) ones of Southern Africa. Burundi's Lake Tanganyika is home to the *Mambu-mutu*. Its description is very much a mermaid, half-human, half-catfish, and prone to lure humans into the water. There the similarities end, for the *mambu-mutu*'s goal is to drown its victims and feed on the blood. Karl Shuker suggests that the source of the *mambu-mutu* is a pod of African manatees, transformed by Native superstition into a vampiric mermaid. Still, the scavenger origin of primates may suggest that some oreopithecine descendants chose not to give up a food source as easily obtained as that of fellow primates.

Lake Kashiba, four hundred miles southeast of Lake Tanganyika's vampiric mermaid, has an even less savory story. It is home to the *chitapo* of the Lamba-speaking peoples. Lakes Namulolobwe and Nakamwale are also inhabited by *chitapo*. These female water spirits lure people to their deaths and serve as a "disposal system" of abandoned, ill-omened infants. Infants, thieves, and witches are also thrown into Lake Namulolobwe,[26] which is located one hundred and thirty miles north of Lake Kashiba. All three lakes are supposed to be connected to other lakes by subterranean waterways, so when anthropologist Brian Siegal argues the *chitapo* did not become a mermaid until the unification of regional water spirits as Mami Wata,[27] he must also admit the *chitapo* is never described. And with potentially hundreds of miles of underwater passages, it is possible the *chitapo/* ape can swim in the underwater routes and remain a phantom to the surface dwellers. They could swim along the passageways, stopping at underground air supplies and dragging human remains with them to scavenge later. This is a trick learned from crocodiles.

In *The Thousand and One Nights*, the Sultana Scheherazade tells stories to stall the Sultan from killing her, as he does all the women he marries.[28] The book is a framing device to compile tales from Indian, Persian, and Arabian folklore that had been retold for centuries. So there may be a kernel of truth buried within the fairy tale.

Two tales prominently feature merbeings. These include "Julanar the Mermaid and Her Son Badar Basim of Persia" and "Abdullah the Fisherman and Abdullah the Merman." Both date back to the 12th or 13th centuries. "Julanar the Mermaid" is a prologue to the adventures of her son. She is a mermaid captured and enslaved but eventually marries the king. Her son must pursue a wife and peace to prove his worth but is first baptized by his aquatic family before setting out.[29] The stories include a commitment to cross-cultural cooperation, particularly evident in "Abdullah."[30] The fisherman and the merman, both Muslims, develop a friendship to the point where the merman brings the fisherman to visit his underwater community. Finally, the two part ways because of a minor theological disagreement over whether Muslims should mourn a death or celebrate the deceased's life. They are considered fairy tales, but they also assume a degree of communication is possible between the two lines of primates.

The Franciscan friar André Thevet is best remembered for his early observations as the cleric on an early colonization attempt in Brazil, but his 1557 travelogue includes material from his visits to the Near East, where he acquired a great deal of secondhand information.[31] Although Thevet claims he interacted with merbeings in Abyssinia (modern-day Djibouti), it could be an earlier account Thevet included in his book.

According to Thevet, "a sea monster having the shape of a man" was stranded on the shore by a flood. His mermaid mate remained offshore, "crying aloud and sorrowing for the absence of her mate."[32] This is a rare case where location is more important than the details. This sighting occurred near the Bab el-Mandeb Strait, connecting the Red Sea with the Gulf of Aden and the Arabian Sea. It is not the only mermaid sighting in the vicinity.

Christoph Fürer von Haimendorff was a Nuremberg nobleman who spent several years exploring the Holy Land. On November 18, 1565, after a visit to Mount Sinai, his party arrived in El Tor on the southwestern coast of the Sinai Peninsula. He found the Red Sea port town to be drab. It had a functional, unremarkable harbor where ships docked with spices from Abyssinia and India. The only thing he found notable was the discovery of a mermaid's skin. He carefully noted the lower part was a fishtail. From the navel up, the creature was human. Only the torso, with breasts, remained. The head and arms had been lost over time.[33] With nothing else of interest, the party continued to their next destination. The missing head and arms suggest the "skin" could be a water ape's discarded protective gear.

Nearly three hundred and fifty years later, Scott Latham, a petty officer aboard the freighter *Stotzenfels,* saw a mermaid in the same area. The southern Red Sea has a spectacular display of bioluminescent plankton, which Latham was studying when the mermaid appeared. The creature disappeared before he could call other witnesses. He was so convinced of the mermaid's existence that he was willing to risk ridicule by relaying his sighting to a reporter when in New York. The account was picked up nationally with surprisingly little overt mocking.[34]

Early in the 18th century, naturalist Benoît de Maillet wrote *Telliamed*. De Maillet was a well-traveled French diplomat, natural historian, and the French consul general at Cairo. He was also a proto-evolutionary biologist who developed a theory of evolution over a century before Charles Darwin published in 1858. *Telliamed* was his theory that all life originated in the ocean, evolving into land-based plants and animals as the waters subsided. According to de Maillett's timeline, mermaids and tritons were not only genuine, but they were also humanity's ancestors. His evidence included sightings of merbeings, such as a 6th century report of "sea people" sightings in the Nile in Lower Egypt.[35] This, according to Shalaby, was an *al-naddaha*, a name given specifically to sirens of the Nile Delta.

They are considered particularly dangerous and usually appear fully human.[36] The witnesses, including a Byzantine officer, observed two merfolk for two hours. The merman appeared threatening with a "fierce air." His hair was red and somewhat bristly. His skin was a shade of brown that the Byzantine felt was similar to that of the Egyptians. Alternately, the mermaid looked "sweet and mild." Her hair was black and shoulder-length. Her skin was white and she had prominent breasts.

In 2009, a mermaid was spotted on Kiryat Yam beach in Israel. The mermaid, "a cross between a fish and a young girl," appeared for several weeks at sunset.

"I was with friends when suddenly we saw a woman laying on the sand in a weird way," a witness said. "At first I thought she was just another sunbather, but when we approached she jumped into the water and disappeared. We were all in shock because we saw she had a tail."[37]

This mermaid makes me suspicious. Much like the 1967 mermaid in Vancouver discussed in the preface, a reward was offered for the capture or photographic proof of the mermaid. And the reward was offered by an organization that would benefit from the resulting publicity. In Kiryat Yam, it was the local Tourist Board.

When the Europeans enslaved Africans, *Mami Wata*'s stories accompanied them. Across the Caribbean, *Mami Wata* again transformed, becoming an amalgamation of *Mami Wata*, South American native water spirits, and the saints of the Catholic missionaries. The islands all have slightly different names for the mermaid/spirit, but Mother Water remains readily identifiable as a water spirit, if not a water ape.

ENDNOTES

1 Bernard, "Ecological Implications of Water Spirit Beliefs in Southern Africa" in *Science and Stewardship to Protect and Sustain Wilderness Values* (2003), 148–154.

2 Bernard, "Ecological Implications," 150.

3 Drewal, "Interpretation, Invention, and Re-presentation in the Worship of Mami Wata." *Journal of Folklore Research,* 104.

4 Cotter, "From the Deep, a Diva with Many Faces." *The New York Times,* 2 April 2009.

5 Drewal, 133-34 fn2.

6 Pontoppidan, *The Natural History of Norway* (1755) II,188.

7 Dapper, *Naukeurige Beschrijvinge der Afrikaensche Gewesten* (1668), 605.

8 Osoba, *Benin Folklore* (1993), 40.

9 Aschwanden, *Karanga mythology* (1989),198.

10 Ittmann, Johannes. "Der Kultische Geheimbund Djĕngú an Der Kameruner Küste," *Anthropos,* 141.

11 "Mermaids stopping Govt work: Sipepa Nkomo." *The Herald* (Harare, Zimbabwe), 30 January 2012.

12 "Mystery surrounds boat tragedy," *The Herald* (Harare, Zimbabwe), 28 September 2010.

13 "Zimbabwe mermaids appeased at pumphouse," *news24,* 12 February 2012.

14 Bernard, "Mermaids, Snakes and the Spirits of the Water in Southern Africa" as quoted in Varner, *Creatures in the Mist* (2007), 16.

15 Pekeur, "Mysterious 'Mermaid' Rises from the River," *Independent Online News,* January 16, 2008.

16 Alexander, *Narrative of a Voyage of Observation among the Colonies of Western Africa,* v.II (1874), 316, plate III.

17 Orpen, "A Glimpse into the Mythology of the Maluti Bushmen," *The Cape Monthly Magazine* (July 1874). The rock painting also appears in How, *The Mountain Bushmen of Basutoland* (1970).

18 Leeuwenburg, "A Bushman Legend from the George District," *The South African Archaeological Bulletin* (December 1970), 145-146.

19 "Chipata Overspill Residents Burn Down Church After Suspecting a Mermaid Inside," *The Zambian Observer,* 13 January 2020.

20 Page, "San Mermaids in Lesotho," *South African Archaeological Bulletin* (June 1978), 2–3.

21 Shalaby, "The Middle Eastern Mermaid" in *Scaled for Success* (2018), 7–8.

22 *Al-Ahmadi, "Houryaat al-Bahr Bayin al-Khoraafa wa al-Asl,"* *Al-Riyadh,* 12 May 2005, quoted *by* Shalaby, 11.

23 de Maillet, *Telliamed* (1750), 231-32.

24 Shalaby, (author's translation), 11.

25 Puckler Muskau, *Egypt Under Mehemet Ali.* volume II (1845), 136–37.

26 Verbeek, *Filiation et Usurpation* (1987), 101.

27 Siegel, "Water Spirits and Mermaids" in *Sacred Waters: Art for Mami Wata* (2008), 303–312.

28 *The Book of the Thousand Nights and a Night* (1897).

29 "Julnar the Sea-Born and Her Son King Badr Basim of Persia," volume 6, 54–95.

30 "Abdullah the Fisherman and Abdullah the Merman," volume 7, 237–258.

31 Thevet, *Les Singularitez de la France Antarctique* (1557).

32 Thevet, The Nevv Found World; or, Antarctike (1568), 28.

33 Füreri ab Haimendorf, *Itinerarium Aegypti, Arabiae, Palaestinae, Syriae, aliarumque Regionum Orientalium* (1620), 40.

34 "What! Saw a Real Mermaid?" *Norfolk (NE) News-Journal,* 13 January 1911. Distributed nationally.

35 de Maillet, *Telliamed* (1750), 230–231.

36 Shalaby, 7.

37 "'Mermaid' Spotted on Kiryat Yam, Beach," *Israel Hayom,* 12 August 2009.

Chapter Nine
Oceania

OCEANIA IS A GEOGRAPHIC REGION that includes Australia, New Zealand, and the areas known as Micronesia, the Philippines, Polynesia, and Melanesia. The area contains thousands of small islands spread across the western Pacific Ocean. You might think that mermaid sightings would be especially prevalent in these areas. But in her book *Mermaids: The Myths, Legends, & Lore*, mythology researcher Skye Alexander explains that sightings of other water monsters are more frequent.

> You'd expect the South Seas from Hawaii to New Zealand to be swarming with mermaids—those lush, tropical islands, coral reefs, and sparkling blue waters seem like the perfect place for merfolk to frolic. Oddly, that's not the case. What we do find, though, are stories of water gods and goddesses—some with fish, snake, lizard, or crocodile appendages. These aquatic deities bear similarities to merfolk and water spirits in other parts of the world—they create and destroy life, they change themselves into people when they want to walk on land, and sometimes they marry human beings. As the South Sea islanders sailed from place to place, they took their legends with them. Over the centuries, their folktales have mixed, morphed, and emerged as delightfully diverse as the people who tell them.[1]

Philippine mermaid encounters are few and far between. And if the mermaid story sounds like a fairy tale, it is probably because it came into the culture during the four hundred years of colonial rule, primarily in Spain and the United States. Even the word for mermaid, *sirena*, is a linguistic vestige of Spanish rule.

In 1978, the Manilla office of the French news service AFP ran a short article.

> In the summer of 1978, Filipino fisherman Jacinto Fetalvero let slip the secret of his recent fishing success. One moonlit night he had met a beautiful mermaid, with "amiable bluish eyes, reddish cheeks, and green scales on her tail." She helped him "secure a bountiful catch.[2]

It was picked up by a few newspapers as a whimsical filler piece. It quickly disappeared until 1993, when Fortean author Jerome Clark reprinted the account in his *Encyclopedia of Strange and Unexplained Physical Phenomena*. Clark could not find further details on Fetalvero and concluded the media coverage brought ridicule, and the fisherman simply refused to discuss the matter again.[3] Clark reprinted the account in his 2012 book *Unexplained!*[4] And the forgotten little encounter of a mermaid helping a fisherman became a part of the permanent Fortean record.

There is a good reason why the Fetalvero mermaid sounds almost archetypal in its content. H. Otley Beyer, considered the father of Philippine anthropology, amassed a vast collection of research, including hundreds of volumes of typescripts compiled by Beyer on the ethnography of the Philippines. His great-granddaughter Charity Beyer-Bagatsing reviewed the typescripts and discovered over forty *sirena* stories. She was intrigued by the fact most of the stories were from the Ilocos section of the island of Luzon. She asked her Ilokana grandmother why the accounts were concentrated among the Ilokano-speaking group while barely mentioned in the rest of the Philippines.

The reply was enlightening. Beyer-Bagatsing's grandmother explained that the rivers in the Ilocos region appear calm on the surface but hide strong currents and undertows. "I heard accounts of swimmers drowning or in a half-dazed state after escaping the grip of an unknown force that tried to drag them to the bottom of the river. Having no other explanation for this phenomenon, the early Spaniards told the natives it might have been a mermaid or *sirena*."[5]

This makes sense. If the mermaids were regular visitors to the Philippines, they would likely appear just as frequently in the other languages of the islands. The concept correlates to French folklorist Michel Meurger's ideas (mentioned in Chapter Five),[6] suggesting that some legends of Canadian water monsters were originally Native stories meant to warn people about dangerous bodies of water. The *sirena* sightings, like Canadian lake monster sightings, would not provide a literal description of a cryptid per se, but rather a social construct. This warning has lost its context and is now just a sea monster story.

This is not to say all encounters have been transformed into unrecognizable generic fairytales. And the few encounters that happened before colonization seem to support that.

In 1534, the Spanish novelist, poet, diplomat, and historian Diego Hurtado de Mendoza reported seeing mermen in the Pacific Ocean on two separate occasions. The first time, the crew of a ship just off the Philippines reported seeing "a fish" in "the shape of a sea-man."

The second sighting was slightly more substantial. The crew saw another sea-man near a "deserted and ill-looking island" in Polynesia. He seemed to want to be noticed and was spotted leaping out of the water. The ship cautiously observed him as he swam around the vessel. "He jumped about in the water like a monkey, diving, washing his body with his hands," but keeping a watchful eye on the humans, giving the impression of intelligence. When someone made motions to throw something at him, the merman dove and resurfaced farther away from the ship, keeping it within sight until the ship sailed on.

These sightings first appeared in print in a 1615 history of Spanish exploration in the South Seas. They were essentially forgotten and brought back into the public eye in 1756, when the accounts were added to a French history of the French Southern Pacific colonies to bolster their claim to Antarctic. The encounters became more widely known when translated into English ten years later, two hundred and thirty years after the sightings.[7]

The ancestral stories of Indigenous groups spread across northern Australia include female water spirits that are similar in description to the mermaid. The most common name for these is *yawk yawk*, a word used in the Kundjeyhmi language of western Arnhem Land that means "young spirit woman."[8] There are also more specific terms, such as in the Binij Gun-Wok community, where a particular type of *yawk yawk* is associated with rivers and rock pools. This is the *ngalberddjenj* (young girl with a tail like a fish).[9] In a recurring detail among the different Indigenous groups, the mermaid/water spirit often comes ashore by "sprouting legs."

Ingaanjalwurr is a rock shelter located in Manilikarr Country, western Arnhem Land. It is situated at the base of a long, steep slope overlooking Red Lily Lagoon. There are seventy-six painted images of rock art, an atypically large number given the size of the shelter. Four of the human-like figures have tails instead of legs. Researchers at the site believe they compare to the style of the images painted three thousand years ago.[10] The shelter remains in use on the trails, and discussions with the native Traditional Owners include recollections of staying there. They note it was a "good feeding ground," especially for freshwater bream and mussels in Ingaanjalwurr Creek.[11] The image resembles the "Mermaid Frieze" at Ezeljagdspoort, South Africa (see Chapter Eight).

In Arnhem Land, the Burarra people around Maningrida believe there are saltwater and freshwater mermaids. Unlike the freshwater *Yawk yawks*, who are more common in stories, there are also saltwater mermaids, known as *Ji-Merdiwa*, who are considered cruel and unattractive to the point where they cannot be represented in art for fear of offending them.[12]

The Gurindji are indigenous people of the Northern Territory, Australia. In a collection of *yijarni* (true stories) translated from Gurindji to English, we find an account by a man named Ronnie Wavehill; he discusses the *karukany*, a freshwater mermaid.[13] Like the European mermaid, the *karukany* of northern Australia is human above the waist and a fish below it. There is a significant difference, and it is an important difference that is exciting to our water ape identification. The Gurindji version of a mermaid can remove its tail and walk on human legs as it pleases. The Wugularr Aboriginal Community of Central Arnhem Land makes the same observation, that "if you pull off their tail they've got legs inside."[14]

Wavehill tells the story of a hunter who visits a river and sees two dark-skinned women sunning themselves among the freshwater crocodiles. In Chapter One, we discussed the misplaced alligator in North America as a possible mermaid/water apeman indicator. Wavehill's story indicates the same symbiotic relationship on the far side of the globe.

Wavehill continues his story with explanations of how the hunter is detected a long way off by his scent, and the mermaids flee into the water. He returns later and stays upwind, slipping into the water and camouflaging himself with weeds. He approaches the *karukayns* and captures one. He catches one (for what will be a very brief marriage) and is able to remove the tail "just like clothes."

Australian freshwater mermaids are not limited to fishtails. Others describe them as half-snake or crocodile,[15] suggesting that among these merbeings, the speed and maneuverability of the fishtail has become less essential for survival and may be more of an aesthetic choice.

These accounts tell us something we have suspected all along in our research. The fishtail is detachable. Wavehill repeatedly mentions the tail allowed the mermaids to move quickly through the water "like a fish."

Recent scientific studies examine the adaptability of the human eye to see underwater among the nomadic sea people of Indonesia.[16] The curved cornea is an adaptation for seeing in the atmosphere with less visual acuity underwater. Among the sea people, they have adapted with pupils that constrict to mere pinholes. This may support Alister Hardy's original 1960 "waterside hypotheses" discussed in Chapter One.

The other way to improve underwater vision is found in many semi-aquatic mammals and birds. If the cornea flattens, it improves underwater vision but at the cost of farsightedness above the water. Returning to Wavehill's freshwater mermaid, the mermaids had a keen sense of smell, but their vision wasn't good enough to detect weeds moving toward them. The mermaids were farsighted!

The Māori are the Indigenous People of mainland New Zealand. The Māori were settlers from what is now called French Polynesia. They arrived roughly in 1320–1350 AD from the Tuamotu Archipelago. They brought with them the *maraki-hau*. It is a half-human, half-fish figure whose effigy is seen on the *epa*, the carved posts on the fronts of many meeting houses.

Historian Gilbert Archey of the Auckland Institute believed the *maraki-hau* evolved from a more common Māori carving where one leg of a human is elongated and often curled.[17] The difference is that the New Zealand version became more elaborate and showed specific characteristics demonstrating that the *maraki-hau* is a marine creature. McLintock's *Encyclopaedia of New Zealand* offered that

The Marakihau are sea denizens who feature in the mythology of the Maori. From the *Dominion Museum Bulletin*, 2, p. 71, 1908.

by the time Archey published his work, the *maraki-hau* had become less elaborate, still "with the characteristic fishtail, but lacking all of the other distinguishing marks."[18] The most common distinguishing mark is a long, tubular tongue of extensive length that ends in a bell-shaped cup or funnel, used to capture fish by sucking them in.

The origin and evolution of the design seem to cause scholars to lose focus on what a *maraki-hau* represents. This is a two-legged creature that has a fishtail. The tongue is a long tube called a *ngongo* in Māori. This is not the word for "tongue" but translates to "suck" or "drink." While scholars argue about the icon's origins, I see a water ape wearing the protective fishtail and using a snorkel to "suck in" air while hunting fish.

But the most compelling evidence rests at the Puke Ariki museum in Taranaki. A house post was discovered in Waitara, Taranaki, in 1919. The figure at the top is a *maraki-hau*. But what makes this *epa* remarkable is that the *maraki-hau* has a bulbous head, three fingers, and a serpentine body. It is interpreted as the Māori view that humans were amphibious in origin. Since *Homo sapiens* was a coastal swelling primate, the Māori are not wrong. The *maraki-hau* fish tail, combined with the non-human head and three-fingered hands, sounds like a hybrid of two different species of water apes: the saltwater mermaid and the freshwater Lizardman. In fact, a bulbous head and three-fingered hands sound remarkably like the Bishopville Lizardman we'll explore in Chapter Eleven. This is also a legend in Micronesia about a Lizardman in the village of Dugor,[19] suggesting Oceania is a hotbed for future research on historical accounts of the various water apes.

A popular story across Micronesia also casually includes the removal of the tail apparatus to free the legs for mobility on land. A version from Ulithi Atoll near the island of Yap is typical.[20] Two porpoise women come ashore at night to watch the islanders dance. They remove their tails, hide them, and walk into the village. After one such visit, a man from the village spies on the women reattaching their tails and submerging. The next time they return to watch the dancing, the man steals one of the tails and hides it. Trapped on land, the woman left behind marries the man, unaware he is the one who hid her tail. Several years later, when the man is out fishing, the wife is cleaning the house and finds her mermaid tail hidden away. She realizes her fin hadn't been lost and knows who is responsible for her being trapped on land. Some versions of the story specifically mention having to soak the tail in water to rehydrate it. She then puts on the tail, goes back to the sea, and never returns to Ulithi or the husband who betrayed her.

In 1979-80, anthropologist Roy Wagner was collecting folktales on New Ireland's central coast when he encountered tales of the *ri*. New Ireland, or Latangai, is a large island in Papua New Guinea. It is home to an aquatic creature with a description resembling a mermaid. It is well known by natives of Central New Ireland. The *ri*, as they are called in Barok, are known as the *ilkai* in Susurunga, and *pishmeri* in Pidgin. Since the natives were adamant the *ri* was a living creature, not a forest spirit, it was outside Wagner's focus. But he investigated and saw something from a distance. It was long and dark. It would surface and then dive back down. Wagner wrote down the information. When the International Society of Cryptozoology (ISC) was formed in 1982, Wagner sent their journal an article about his mystery animal.[21]

He interviewed local fishermen whose descriptions of the creature varied, but they agreed on several points. The upper portion of the body was human-like, with long, dark hair in both sexes. The *ri's* skin color was light brown, and gender was easily discernable by genitalia. Its fingernails were long and sharp, and the palms of the hands were deeply ridged and calloused. Witnesses all agreed there was something "different" about the *ri's* mouth but couldn't be more specific.[22]

The ISC was fascinated by the possibility of a large unknown aquatic mammal. The following year, Wagner joined *Cryptozoology* editor Richard Greenwell on a small three-person expedition to New Ireland to investigate.[23] They discovered that in the villages further north in central New Ireland, the dugong and the *ri* are considered to be the same animal. But in Ramat Bay and further south, the natives explicitly differentiated between dugong and *ri* as two distinctive animals.

They heard that villagers at a hard-to-reach place called Nokon Bay spotted *ri* almost daily. Once there, they discovered the *ri* appearances had a pattern. The animal came into the bay, presumably to feed, at dawn and dusk. At one point, Wagner and Greenwell were able to maneuver their dinghy to within fifty feet of the *ri*, but the creature dived and never reappeared. Wagner's photographs were "murky, poorly focused ones that show the tail of an unidentifiable marine animal raised out of the water."[24]

A second expedition was arranged.[25] In February 1985, a dozen members of the Ecosophical Research Association arrived to conduct a far more advanced operation. A boat was chartered specifically for diving operations. It came equipped with radar, side-scan sonar, satellite navigation, and other technical amenities. This report stated conclusively that the *ri* was a dugong. They had photographed dugongs surfacing in the mornings and evenings, as Wagner had first reported, and photographed the creature feeding underwater. The ISC disappointedly closed the file of the *ri*.

However, in reviewing this case, two things are important to note. First, the description of Wagner is not that different from sightings on Morotai Island in Indonesia that we discussed in Chapter Seven.

Second, and more significantly, the 1985 researchers used a boat with motors, which they mentioned several times in their report. Wagner's first sightings were from the shore. The second trip used a dinghy, which used paddles. It seems an oversight that no one considered whether merbeings had accompanied the dugongs into the bay. The noise of the engines may have scared off the merfolk, leaving only the dugongs to be observed by the noisy researchers. With this in mind, it may be unlikely that New Ireland can be wholly dismissed as a potential water ape habitat.

Australia offers other insights. In his *Following the Equator*, author Mark Twain tells of a time he was on a boat in the evening, fifty miles out of Sydney, Australia.

> Presently, a quarter of a mile away you would see a blinding splash or explosion of light on the water—a flash so sudden and so astonishingly brilliant that it would make you catch your breath; then that blotch of light would instantly extend itself and take the corkscrew shape and imposing length of the fabled sea-serpent, with every curve of its body and the "break" spreading away from its head, and the wake following behind its tail clothed in a fierce splendor of living fire. And my, but it was coming at a lightning gait! Almost before you could think, this monster of light, fifty feet long, would go flaming and storming by, and suddenly disappear.
>
> And out in the distance whence he came you would see another flash; and another and another and another, and see them turn into sea-serpents on the instant; and once sixteen flashed up at the same time and came tearing towards us, a swarm of wiggling curves, a moving conflagration, a vision of bewildering beauty, a spectacle of fire and energy whose equal the most of those people will not see again until after they are dead.[26]

He explains it was not an "electric sea serpent." Instead, it was "porpoises aglow with phosphorescent light." The ship was passing Jervis Bay as a dolphin pod passed. Their passage disturbed the bioluminescent plankton that the area is known for. The glowing dolphins suggest another possible explanation for the reports of bioluminescence in mermaid sightings. Could our aquatically adapted apes glow because they are coming ashore and are coated with glowing algae?

Apes on land developed decreased reliance on sense of smell as visual cues became more important; detecting fruit by scent is less effective than seeing the color change in ripe fruit.[27] To *Oreopithecus* descendants, underwater vision was vital, limiting their sight out of the water and making them more dependent on scent to avoid predators. That heightened sense of smell that diminished for all other apes and monkeys (including humans) did not lessen among oreopithecine branches. This indicates that it may be difficult to see a mermaid because they can detect you from a great distance by scent. This also raises an interesting sidebar about the relic hominids known as the Neo-Giants, more familiarly known as Bigfoot and the Sasquatch.

In his book on scents, paranormal researcher Joshua Cutchin devotes sixty pages to discussing the variety of foul odors associated with hominid encounters.[28] He is noncommittal about why so many hominid reports include an olfactory component. Cutchin offers four different potential rationales for the stench, ranging from environmental camouflage to deliberately triggered sweat glands used as a defensive mechanism. The one rationale he doesn't include is that mammals have a unique, genetically determined scent called an odortype.[29]

Odortypes distinguish one individual from another member of the species. They are usually associated with urine, but the odorants are also in blood in masked form. An ape that evolved in the swamps must have a particularly potent odortype to compensate for the swamp's background odors and their diminished above-water vision. Indeed, an oreopithecine water ape might have seconds to determine a friend from a foe and stay or flee accordingly. It may be that the odortype is so complex that a primate, one that was not losing its scent in favor of color vision, might detect all sorts of information: tribal affiliation, recent whereabouts, recent food sources, or ovulation of a potential mate. The loss of scent in favor of vision can be documented genetically.[30] But the testing doesn't include the cryptid apes because science doesn't believe they exist.

ENDNOTES

1 Alexander, *Mermaids: The Myths, Legends, & Lore* (169), 2012.
2 "Half-human-half-fish Beauty," *Eastern Horizon* 17 (9), September 1978, page 51.
3 Clark, *Encyclopedia of Strange and Unexplained Physical Phenomena* (1993), 212.
4 Clark, *Unexplained!* (1998), 464.
5 Pacita Malabad Beyer interview in Bacchilega and Brown, *The Penguin Book of Mermaids* (2019), 213.
6 Meurger, *Lake Monster Traditions* (1988).
7 Brosses, "Article VIII—Diego Hurtado to Polynesia." *Terra Australis Cognita*, vol I. trans John Callander (1766), 123–4.
8 Hayward, "Swimming Ashore" in *Scaled for Success* (2018), 189.
9 Green, *Togart Contemporary Art Award* (2007), 20.
10 May, et al, "The Rock Art of Ingaanjalwurr, Western Arnhem Land, Australia," in *The Archaeology of Rock Art in Western Arnhem Land, Australia* (2017), 54–56.
11 May (2017), 62.
12 Whitfeld, "Mermaid tales appear in myths around the world — Arnhem Land included," *ABC RN Late Night Live*, 10 July 2018.
13 Wavehill, "Karukany (Mermaids)" In *Yijarni: True Stories from Gurindji Country* (2016), 13–20. Excerptedim Bacchilega and MBrown, eds. *The Penguin Book of Mermaids* (2019).
14 Wugularr Aboriginal Community with Liz Thompson, *The Mermaid and Serpent* (2010), 5.
15 Alexander, 181.
16 Gislén and Schagatay, "Superior Underwater Vision Shows Unexpected Adaptability of the Human Eye," in *Was Man More Aquatic in the Past?* (2018), 164–172.
17 Archey, "Evolution of Certain Maori Carving Patterns," *Journal of the Polynesian Society* 42 (3) (September 1933), 178.
18 McLintock, *An Encyclopaedia of New Zealand* (1966), v2, 411.
19 Winkler, *Stories of the Southern Sea* (2016), 214–15.
20 Poignant, *Oceanic Mythology* (1967), 82.
21 Wagner, "The Ri—Unidentified Aquatic Animals of New Ireland, Papua New Guinea," *Cryptozoology* 1 (1982), 33–39.
22 Wagner (1982), 37–38.
23 Wagner, "Further Investigations Into The Biological And Cultural Affinities of The Ri," *Cryptozoology* 2 (1983), 113–125.
24 Smith, "Cryptozoology: Seeking Mermaids, Living Legends, Mythical Monsters." *Chicago Tribune*, 27 January 1985.
25 Williams, "Identification of the Ri Through Further Fieldwork in New Ireland, Papua New Guinea," *Cryptozoology* 4 (1985), 61–68.
26 Twain, *Following the Equator* (1897), 109–10.
27 Dominy and Lucas, "Ecological importance of trichromatic vision to primates," *Nature* (2001), 363–366.
28 Cutchin, *The Brimstone Deceit* (2016), 183–242.
29 Yamazaki, et al, "Odortypes: Their Origin and Composition," Proceedings of the National Academy of Sciences (1999), 1522–1525.
30 Gilad, et al, "Loss of olfactory receptor genes coincides with the acquisition of full trichromatic vision in primates." *PLoS Biology* (2004) E5.

Chapter Ten
The Mysterious Ones

THE VARIETY FOUND IN THE EVOLUTION of merbeings is not at all a surprising development. If they are indeed the offspring of *Oreopithecus*, these creatures have been around for at least nine million years and have been adapting to their environments the entire time. We have the example of human beings developing from a common ancestor many millions of years ago into several surviving varieties, such as *Homo sapiens*, *Homo gardarensis*, *Homo neandertalensis*, and *Homo erectus*. In the future, upon physical scrutiny, it could be discovered that the water apes might have diversified into many forms.

One of the varieties already evident in historical accounts are pygmy water apes. These are diminutive creatures with sleek, dark bodies, luminous eyes, and an affinity for water habitats. Again, they have been tagged with many names. Many peoples have encountered them and awarded them a name and a place in their conceptions of the local natural world. These small creatures are seen as part of the environment while remaining distant from human populations. Logic would tell us that they are concerned with their own survival, and keeping humans at a distance is integral to their longevity.

Belgian-French zoologist Bernard Heuvelmans believed the majority of these "proto-pygmies" were found in Africa.[1] Biologist Ivan T. Sanderson disagreed, believing these proto-pygmies were tropical creatures with global distribution but was willing to consider a broader range.[2] There exist "Little People" accounts across the globe, regardless of how hospitable the climate is. Isolation from humanity takes precedence. It is interesting to note that the more nature-oriented cultures have most of the interactions with the Mysterious Ones. For example, John Bierhorst, folklorist and translator of over thirty books on lore, hints that records of rare contacts did occur in Europe, but the accounts are hidden in European fairy tales.

In 1830, on the tiny Hebridean island of Bennecula, people were cutting seaweed on the shore of Culla Bay. One woman heard a splash in the water and noticed a creature in the form of a woman in miniature, only a few feet away. Alarmed, the woman called to her friends, and the others rushed to the spot. The creature did somersaults and swam about in various directions. Some men waded into the water to seize her, but she was too fast. Some boys threw stones at her, one of which struck her in the back. A few days later, the creature was found dead at Cuile, Nunton, nearly two miles away.

> The upper portion of the creature was about the size of a well-fed child of three or four years of age, with an abnormally developed breast. The hair was long, dark, and glossy, while the skin was white, soft, and tender. The lower part of the body was like a salmon, but without scales. Crowds of people, some from long distances, came to see this strange animal, and all were unanimous in the opinion that they had gazed on the mermaid at last.[3]

The Benbecula Mermaid story was first recorded in the *Carmina Gadelica*, a multivolume book of Scottish prayers and folktales compiled in 1900, which also noted that the local sheriff ordered a coffin be made for the mermaid. She was buried on the shore near where she was first seen, and there were (as of 1900) "persons still living who saw and touched this curious creature, and who [gave] graphic descriptions of its appearance." This is also one of the few cases where an official but unsuccessful effort was made to relocate the burial site in 1994.[4]

In 1823, a mermaid became entangled in fishing lines off Yell in the Shetland Islands.

> The statement is, that the animal was about three feet long, the upper part of the body resembling the human, with protuberant mammae like a woman; the face, forehead, and neck, were short, and resembling those of a monkey ; the arms, which were small, were kept folded across the breast; the fingers were distinct, not webbed; a

few stiff long bristles were on the top of the head, extending down to the shoulders, and them it could erect and depress at pleasure, something like a crest. The inferior part of the body was like a fish. The skin was smooth, and of a grey colour. It offered no resistance, nor attempted to bite, but uttered a low plaintive sound. The crew, six in number, took it within their boat, but superstition getting the better of curiosity, they carefully disentangled it from the lines, and a hook which had accidentally fastened in its body, and returned it to its native element. It instantly dived, descending in a perpendicular direction.[5]

Additional details came to light after the encounter was reported by Robert Hamilton, a professor of natural history at Edinburgh University.

They had the animal for three hours within the boat; the body was without scales or hair; was of a silvery grey colour above, and white below, like the human skin; no gills were observed; nor fins on the back or belly. The tail was like that of the dog-fish; the mammae were about as large as those of a woman; the mouth and lips were very distinct, and resembled the human.[6]

Accounts from American Native peoples show the Mysterious Ones were far more widely distributed than Heuvelmans had initially believed, and there are similarities in descriptions of the creatures across regions.

Although the stories are different from one region to the next, there is a sameness that cannot escape notice. Certain characteristics keep turning up, and it would seem that the little people have a style of life all their own whether they are imagined by the Inuit or the Cherokee, the Zuni or the Iroquois.

Their physical height at its upper limit is about four and a half feet, with two to four feet as the typical range (though some storytellers, fancifully, imagine them much smaller). The little people are wilderness dwellers seldom seen by average-sized humans. Yet, when visited, they prove to be generous hosts. They have inexhaustible food supplies, which they share or withhold as they see fit. They are capable of mischief, yet they can bring good luck. As small as children, they are wiser than adults. And though they are said to be old, even extremely old, they have great bodily strength.

In part, at least, these qualities suggest the folkloric little people of Europe. In the Grimms' fairy tale "Snow White," the seven little wilderness dwellers with their hospitable invitation—"Stay with us and you shall want for nothing"—offer much that could recommend them to a Native American audience. It might even seem that the idea of little people itself had been borrowed from the Old World-like "Snow White" in Native American tradition, and in general the differences between the folkloric little people of Europe and America are as great as the similarities.[7]

Researchers must take to heart the declaration of Deskaheh, a famous leader among the Iroquois. When approached by a folklorist who was seeking knowledge about the Little People, Deskaheh had some excellent advice. He suggested that the Natives had been in America for so long and in such intimate contact with their natural environment that they had formed relationships with beings and things Europeans had no knowledge of.[8]

Slowly we have been learning the wisdom in this observation. From the giant birds they called Thunderbirds and Great Owls to the hairy giants popularly known as Bigfoot, we have seen these things in nature for ourselves. These fabulous animals have a place in the modern world because they are real and will not conveniently disappear. Even the Little People now have a scientific name (*Homo floresiensis*). Likewise, they will be demanding recognition as a fundamental part of nature, just as the Iroquois said they were.

Another name given to them in North America is "The Mysterious Ones." That name captures the essence of these creatures for all of us today. They are mysterious and will remain so for a long time, even if we were to take an interest in their reality and attempt to learn more about them. So they might well be known to ethnic groups while remaining little known at the same time. So, for now, we must be content with a brief look at their rare appearances and some notices of them taken by ethnographers and chroniclers of the curious.

Ornithologist Katharine Scherman recalls a bird-watching expedition to Bylot Island, four hundred and fifty miles north of the Arctic Circle off the northeastern coast of Baffin Island. Their guide was ldlouk of the Tunnunermiut Inuits. As they climbed inland mountains, they noticed an Ice Age shoreline, now far above sea level.

> Scattered over the Midland were small uneven rings of stones. We guessed that these were caused by frost action, but asked ldlouk. He frowned at them, the frown that meant this was going to be difficult for him to explain in his combination language. Then his face cleared. Little people, he said with his hands. Littler than children. People who ate only rabbits. Down at Canada Point, the western corner of Bylot Island, there was a whole camp of little rent rings with rabbit bones in them. But there were no Little People any more.[9]

This area may prove to be essential to understanding the Mysterious Ones. To the south, off the southwestern shore of Baffin Island, Luke Foxe, an English explorer, sought a northwest passage in July of 1635. He reported that he came across an island used to inter the dead in stone cairns topped with nine-foot long, four-inch-thick wooden planks. They robbed the graves, mainly to use the boards as firewood. They observed the remains were wrapped in deer skins, and the bodies were positioned to the west, buried with their bows and arrows, bone lances, and other objects. Most notable was that the tallest corpse was under four-feet long.[10] Does this cemetery demonstrate the arctic Mysterious Ones were assimilating Inuit customs, or was there a calamity, and the Inuits who came across the scene buried the remains in their tradition? Our lack of knowledge remains the problem.

On July 10, 1873, Rebecca Young took her seven-year-old son Freddie and several of his friends to the beach in Brewster, Massachusetts. They came across something on the sand when they arrived on the Cape Cod beach. The head resembled a child, while the rest of the body was a fish. Mrs. Young was frightened, but seven-year-old boys are fearless, and they wanted to know if it was dead or alive. So they threw sand into its eyes. The merbeing wailed like a child and rolled across the beach and into the water. It swam away, "keeping its head above the surface and resembling in every manner that of a child swimming."[11] Brewster residents were divided about whether the creature was left by the tide or if it came ashore overnight on its own.

On December 30, 1879, the *Lewes Breakwater Light* reported that a body the size of a six-year-old mermaid was found on the beach near Cape Henlopen lighthouse in Delaware. The news quickly spread through newspapers.

> On Tuesday of this week Captain Raymond, keeper of Life Saving station No. 3, found on the beach what he supposed to be a mermaid, which had been washed up from the sea. It was dead when it came on the beach and in a slight stage of putrefaction.
>
> Captain Raymond describes it as being about the size of an ordinary 6 year old boy, and to the waist or middle of the body resembled a boy in every particular. He says that his face, head, neck, arms and bust as well as its hair was perfect in appearance to that of a human being. There were no fingers on the hands, but a coarse, moppy hair like the frizzled end of a whalebone, supplied their place. The lower portion of the body from the middle or waist downward resembled that of a shark, the tail being covered with a hairy substance similar to that of the hands.
>
> The sea nymph has created great excitement in the vicinity of the station, and many of the inhabitants thereabouts think its presence forebodes bad luck.
>
> Capt. Fowler says that the 'tarnal critter' came there for no good, and that it betokens a terrible shipwreck and fearful loss of life, which is soon to happen on that part of the coast.[12]

Captain Fowler may have been correct. In 1970, a NOAA hydrographic survey discovered the wreck of the brigantine *DH Bills*, which left Liverpool days after Fowler's prediction and sank during a March blizzard twelve miles off Cape Henlopen.

Such dire predictions aside, very little is malevolent about these creatures. There are exceptions, such as the Little People of the Arapaho, Shoshoni, and neighboring tribes referred to as "cannibal dwarves." The tradition is that these creatures would leave their hearts in camp when they hunted, making them immortal. The method of killing these mortal enemies was simply to find and puncture the hearts.[13] If water apes need protective gear, it is not a stretch to suggest that the original intent was to poke holes in the Mysterious Ones' version of a wetsuit, stranding them on land where they were more easily killed.

In Japan, *kappa* are said to be about the size of a small child, inhabiting the ponds and rivers of Japan.[14] There is a distinction between the *kappa* and *ningyo*. The *ningyo* is considered the Japanese version of a

A flier on a mermaid (Ningyo no zu) caught in what is now Toyama Bay, Japan, 1805. *Wikimedia Commons.*

mermaid (see Chapter Seven). *Kappa* have a circular depression atop their heads filled with water when they leave their aquatic environment. To avoid their mischief, it was advised to make them spill the water, which rendered them powerless. Is this another reference to tampering with protective gear, obfuscated by the centuries?

For over a century, the Taunton River on the South Shore of Massachusetts was not living waters, be it mermaids or fish. Sewerage from the entire central portion of the City of Taunton drained into a tributary of the Mill River. Mill River had been dammed above the town, and the irregular water flow resulted in effluence on the banks and in the bottom mud. The Mill River flowed through the middle of the city and into the Taunton River. In his *Picture-writing of the American Indians*, ethnologist Garrick Mallery, who specialized in Native American sign language and pictographs, mentions an 1886 report of damage to the Dighton Rock petroglyph boulder in the Taunton River. To improve visibility for tourists, local residents were scrubbing the inscriptions with a broom to remove what Mallery politely referred to as "silt" that had accumulated from the daily tidal immersions.[15] Truth be told, the silt was raw sewage and mill run-off.

As time passed, the river began to heal itself from the pollution. And one sign of returning life was a resurgence in reports of the Mysterious Ones called *pukwudgies*. The name is from the Ojibwa and is used across Algonquian-speaking tribes to describe a mischievous dwarf or elf. The further east from the Great Lakes you travel, the more malicious the pranks become in *pukwudgie* folklore. It may not be a coincidence that the farther east you traveled, the more polluted the rivers were.

Most modern *pukwudgie* reports are from the area surrounding the Freetown State Forest in Massachusetts. This region is such a hotbed of paranormal and cryptid phenomena that cryptozoologist Loren Coleman coined the phrase "Bridgewater Triangle" to describe it. And running through the middle of the Bridgewater Triangle is the Taunton River. The *pukwudgies* avoid contact with humans and can literally disappear from view so quickly and effectively as to appear supernatural. The few who see more than a brief glimpse of a *pukwudgie* often report being stalked by the creature, literally terrified into silence.[16] Apparently, they have not forgiven humankind for destroying their habitat.

Less than twenty-five miles from the northern point of the Bridgewater Triangle, another Mysterious One was reported. Unlike the maliciously angry *pukwudgies*, the "Dover Demon" appears to have been relatively harmless. It was so unusual in its appearance as to create a sensation among humans when it turned up along the Charles River within fifteen miles of a heavily populated city. But then, its appearance anywhere is likely to offend those content with the status quo, denying that animals of any unknown kind appear to us today.

The "Dover Demon" was a title that Coleman applied to the creature after interviewing the witnesses. The incidents took place over two days in April of 1977 as various people caught a glimpse of a small creature in areas along the Charles River around Dover. One of the teen witnesses was Bill Bartlett, who was also a skilled artist. His drawing of the creature remains a widely reproduced image in cryptozoology. In Coleman's 2013 book *Monsters of Massachusetts*, the creature is described as follows.

> [T]he creature grabbed onto a rock with long fingers, slowly turned its head toward the approaching car, and stared into the light. . . . The entity possessed an unusually large head shaped like a watermelon and about the same size as its trunk. In the center of its head glowed two large, round, glassy, lidless eyes shining brightly 'like two orange marbles' in the glare of the headlights.[17]

As a well-known expert in cryptids, Coleman was consulted by the newspapers. He observed that the reports of the three-and-a-half to four-foot-tall Dover Demon—a creature with rough, shark-like skin—sounded similar to an earlier encounter.[18] The spindly arms and legs with large hands and feet reminded Coleman of the August 1955 "Hopkinsville goblins" incident in Kentucky, reported initially as a UFO landing.[19]

In a 1929 article about the Mysterious Ones, they're described as "Water Indians."

> Like the Greeks, the Indians had their Lares and Penates, nymphs and dryads. There are the Water Indians with strangely-shaped feet who live in the bottoms of lakes and rivers and who come forth only at night. In the morning footprints and little heaps of pebbles at the water's edge attest their nocturnal exploits. Unless treated with great respect and veneration, these Water Indians may steal a little papoose and take it far down under the water to their own world and never return the child to its bereaved parents.[20]

A brief encounter with a pygmy water ape would explain a similar situation that occurred in Toronto in August of 1977. A man, who has requested to remain anonymous, crawled into a small cave near his Parliament Street apartment. He was looking for a kitten he had been caring for.

> "It was pitch black in there...I saw it with my flashlight. The eyes were orange and red, slanted.... It was long and thin, almost like a monkey...three feet long, large teeth, weighing maybe 30 pounds with slate-grey fur."
> Ernest speaks reluctantly of what happened next. He is convinced the thing spoke to him.
> "I'll never forget it," he said. "It said, 'Go away. Go away.' in a hissing voice. Then it took off down a long tunnel off to the side...I got out of there as fast as I could. I was shaking with fear."[21]

Reporter Lorrie Goldstein went with Ernest to the site of the event. It was between two buildings and had been blocked by a fallen concrete slab since the encounter. A tunnel went back ten feet and then turned. A sewer worker inspected the site just in case it was a safety hazard. He told Goldstein that the tunnel was probably created by erosion from poor drainage. It could possibly lead into the sewer system, but not likely. He didn't discount the encounter, saying odd things happen underground in a cityscape.

These creatures have kept their distance even from the Native Americans. In the 20th century, people encountered the *Tonh-kyanh-hee* in Oklahoma and Texas. The Kiowa knew the creatures as the "People who live under the water and mud." The few who have seen them described them as black, two to three feet tall, with large eyes, and covered with scales.

In Quebec, among the Montagnais and Naskapi, there is an Underwater Man (*Tambe'gwilnu*), considered the equivalent of a merman. *Tambe'gwilnu* is regarded as a guardian or owner of springs. Travelers leave a present of tobacco in gratitude for the privilege of drinking the cold water.[22] This could be related to the Wabanaki reference to *Alambe'gwinosis*, the "underwater little man" of the Penobscot.[23]

A more informative account tells how a horse appeared to kill one of the creatures by dragging it out of a lake in Texas. It was small and black, with hard and scaly skin. Two more creatures came out of the lake and stood over the body. A Kiowa chieftain approached them and communicated with speech and signs. He returned to his people and told them the following.

> He said, they told me they live under the water and the mud and that that is their world, just as we live on land. They rarely come out on land and instead catch what game there is that happens to be near or in the water. When the chieftain asked if they ever caught humans, the creatures said no, and that humans are very hot to their touch. Thus it is that they leave humans strictly alone and therefore there is no reason for us to ever meet, except on rare instances like this one.[24]

On the Bahamian Island of Andros, the *chickcharney* similarly avoids humans. They are said to stand at only three feet tall with three toes on each foot and three visible fingers on each hand. Their red eyes are large and often compared to owls. This comparison to owls has complicated the task of the cryptozoologist, as the extinct Bahamian Giant Barn Owl called the same island home. More than one skeptic has suggested the owl's survival on Andros Island overlapped the arrival of man, making the *chickcharney* a memory of the owl passed down through generations. However, the bones of the Bahamian Giant Barn Owl were not found until 1937.[25] Before that, as late as 1926, *chickcharnies* were considered humanoid, closer to fairies than birds.[26] An even earlier account from 1886 describes these "little people" as tree-dwellers who look "like tiny men covered with hair."[27]

And of particular note, the Bahamian Giant Barn Owl was flightless.[28] The *chickcharney* builds its nests in treetops, often tying several trees together to form a shelter. If the Bahamian Giant Barn Owl, the same size as the pygmy water apes, hunts on the ground, hiding at night in the trees is perfectly logical survival behaviorism for a creature being forced to avoid its watery natural habitat.

This is not the only example of water-adjacent Mysterious Ones. Red Tide, predators, pollution, or any number of reasons could explain the decision to return to land. Among the Indian tribes of California, the "Water Babies" were described as small, dwarf-like men in traditional Native dress. They had long hair and lived along streams and water holes.

Similarly, the Olmec knew of the *chaneques*, described as "old dwarfs with faces of children."[29] The *chaneques* live in waterfalls, hearkening back to the South African mermaid's preference for "living waters."

When the first Polynesians arrived in Hawaii, they found dams, fish ponds, irrigation ditches, and even *heiau*, or temples, that showed no trace of human involvement. The assumption was that this was the work of the *Menehune*, a race of dwarf people in Hawaii.[30] The *Menehune* traditionally live in the deep forests and hidden valleys, not the water. Their association with this water-themed construction could be an attempt to make a new habitat of living water.

The presence of these diminutive Mysterious Ones alongside the other merfolk of North America clearly indicates how complex the answers to the mysteries of the water apes will be. People often want to simplify the neglected issues of this kind, as if centuries of fear and avoidance will not have allowed many things to survive. The lesson we've learned in recent decades is that many people have assumed voluntary blindness to creatures that the Native Americans have seen for centuries. No one paid much attention before, but now we have more reasons to listen and learn.

ENDNOTES

1 Heuvelmans, *On the Track of Unknown Animals* (1959), Chapter 19.

2 Sanderson, Traditions of Submen in Arctic and Subarctic North America, *Pursuit,* 1983.

3 Carmichael, *Carmina Gadelica* Vol. 2 (1900), 305.

4 "The Mermaid's Grave, Nunton, Benbecula (S Uist parish): putative grave-marker," Discovery & Excavation in Scotland, 1994, 95.

5 Hamilton, *Amphibious Carnivora,* [c.1840], 286–288.

6 Hamilton, 287.

7 Bierhorst, *The Deetkatoo* (1998), viii-ix.

8 Witthoft, "Cherokee-Iroquois Little People," *Journal of American Folklore* 59 (234), October-December 1946.

9 Scherman, *Spring on an Arctic Island* (1956), 228.

10 Fox, *North-west Fox* (1635), 202–203.

11 "Brewster," *Provincetown Advocate,* July 16, 1873

12 "A Mermaid at Lewes," *Daily Gazette,* 5 January 1880.

13 Dorsey and Kroeber, *Traditions of the Arapaho* (1903), 122–23.

14 Foster, *The Book of Yōkai* (2015), 157.

15 Mallery, *Picture-writing of the American Indians* (1972 rep), 86–87.

16 Balzano, *Dark Woods: Cults, Crime and the Paranormal in the Freetown State Forest* (2008), 107–111.

17 Coleman, *Monsters of Massachusetts* (2013), Chapter 1.

18 "This monster may be a hoax, then again…" *The Boston Globe,* 16 May 16 1977.

19 "Story of Space-Ship, 12 Little Men Probed Today," *Kentucky New Era,* 22 August 1955.

20 Hall, "The Mystery of the Thunderbird," *Overland Monthly and Out West Magazine,* April 1929, 109–110.

21 Goldstein, "Tunnel monster of Cabbagetown?" *Sunday Star,* 25 March *1979.*

22 Speck, *Naskapi* (1935), 68.

23 Speck, 68fn11.

24 Bates, "Legends of the Kiowa," I *NFO Journal,* May 1987, 4–10.

25 Wetmore, "Bird remains from cave deposits on Great Exuma Island in the Bahamas," *Bulletin of the Museum of Comparative Zoology,* October 1937, 436–439.

26 Moseley, *The Bahamas Handbook,* 1926, 66.

27 Gardiner, "Alligators in the Bahamas," *Science,* 22 October 1886, 369.

28 Marcot, *Owls of Old Forests of the World,* U. S. Department of Agriculture, 1995, 26.

29 Bernal, Ignacio. *The Olmec World* (1969), 100.

30 Baldwin, "Menehune," *Storytelling: An Encyclopedia of Mythology and Folklore* (2015), 2: 303–4.

Chapter Eleven
1988: The Year of the Lizardman

IT IS A NATURAL TENDENCY FOR HUMANS to view themselves, at the beginning of each century, as the most sophisticated of all people in history. With the internet, cellular devices, and virtual reality, we can explore the corners of the globe without leaving the comfort of our sofas. We can all feel pretty sophisticated.

We find similar sentiments expressed early in the 20th century. People felt the world had become very well known by the 1900s. Mysteries had diminished, if not vanished, under the relentless onslaught of humanity's desire to explore, define, and map planet Earth.

However, every few years, this complacent contentment with our sophisticated outlook is jolted by some event that threatens to expose a failure in our textbook versions of things. In 1967, the event that rocked the world was a motion picture of a hairy giant in California. The Patterson-Gimlin film is a clear view of a Neo-Giant in full color. It was a Bigfoot sighting for all of humankind. This new information has changed modern science. By the 2000s, a handful of prominent scientists began to feel they could, without risk of ridicule, come out publicly in support of pursuing the evidence for the Neo-Giants.[1]

In 1977, the event that shook our perceptions was the attack upon a young boy in Illinois by a giant bird known as the "Thunderbird." Before this return to public notice, the huge bird had been confined to legends. In front of several witnesses, adults and children, the victim was picked up and carried at least thirty-five feet before being dropped. The bird flew off, continuing on a path south out of the state, as was indicated by later reports.[2] There was no explanation for the encounter, but there were eyewitnesses to the continued existence of giant birds. These creatures were also claimed to be responsible for attacks that spanned a century.

In 1988 another big event that jarred the American consciousness occurred in South Carolina. People reported seeing a strange creature. Cars were mysteriously damaged. Its tracks were found by professional law enforcement officials. Seeing the beast was scaly and without hair, the eyewitness and locals called it "the Lizardman." The primary event was the experience of a young man who got a good look at the creature as it ran toward him. He escaped by car. The year 1988 became "The Year of the Lizardman."

Lizardman by Bill Rebsamen, from *Mysterious America* by Loren Coleman, Paraview Press, 2001.

The stories of the creature broke into the news in July of 1988. For weeks, around Bishopville in South Carolina's Lee County, it remained a topic of local discussion. Then, the story hit the newswire. In the center of the action was Scape Ore Swamp. The swampy creek flows from the northern part of Lee County, passes west of Bishopville, and then exits the county to join the Black River east of Sumter.

In 1988, something seven feet tall and covered in scales left behind large three-toed footprints near Scape Ore Swamp. The thing chased one young man in the early morning hours of June 29. It was his story that triggered the whole Lizardman saga.

Seventeen-year-old Christopher Davis was changing a tire on his 1976 Toyota Celica when he first saw the thing.

> I had finished changing the tire and was putting the things in the trunk," he said. "The moonlight was out. I turned around and saw a red-eyed devil. He was about 30 yards from me, in the field. It had real long arms. When he would run, his arms would swing.
>
> "I ran to the driver's side and got in. When I was sitting in the car I saw him from the neck down. I pulled off

and after about two yards he jumped on the roof." Davis said he could see the creature's three-fingered hands sticking down from the top of the windshield as he got up to speed.

"I saw hands, rough-looking black-fingernailed hands," Davis said. "After he jumped up on the car he grunted. A deep grunt. He grunted one time." [3]

The thing fell off the car but chased after the vehicle, and although Davis reached speeds up to forty miles per hour, the creature kept up. As the thing matched his speed, Davis realized this creature was not just someone in a monster outfit. He reached his family's home, blew his car horn, and ran inside, leaving the car running. His father, Tommy Davis, recalled how his son had been upset; he told his family he had seen something "seven feet tall, with red eyes, and three fingers on each hand."

Christopher Davis was given a polygraph test on August 18, 1988. A police captain administered the examination in Sumter, NC. When asked specific questions about his story, Davis passed them all. His responses were shown to be truthful. In response to this news, several publications unleashed reflections on how unreliable lie detector tests are thought to be. So, it seemed that only failing the test would have been considered important. But, of course, passing the test didn't make much of a ripple.

Lee County Sheriff Liston Truesdale said that another eyewitness, a Browntown resident, claimed in a sworn statement to have crossed paths with the creature in the fall of 1987. The sheriff's department was told about this earlier incident in June of 1988, but they didn't believe it. Instead, they collected a sworn statement on August 13, 1988, after they began taking the Lizardman seriously. The deputy didn't put much stock in the story until Davis made his claim.

"What adds credibility to his story is that we knew about him before we knew about Chris Davis," the sheriff said. "We've been looking for him but couldn't find him."

In an August 13 statement, George Holloman Jr, a 31-year-old construction worker, said he had ridden a bike to a flowing spring about 12:30 a.m. one day last fall. Holloman drank some water, lit a cigarette turned around and "saw it across the road." Truesdale said, "What he saw, he thought, was a dead tree that had been struck by lightning. Then it moved."

The sheriff said Holloman described it as between 7 and 8 feet tall, and it "stood up like a man," Truesdale said. "A car passed by, and that's when he saw the eyes. He said they glowed when the car passed. "The creature went back in the swamp after the car passed. Holloman told the sheriff. "He thought it was a 'hant,' what rural folks call ghosts, the sheriff said.[4]

Drivers Tom and Mary Waye reported that a creature of incredible strength had pulled the chrome straight off their car during the night on July 14. Four days after the Wayes reported the damage, Sheriff Truesdale found odd tracks that he thought might "could possibly" belong to a bear. He called in biologist Matt Knox of the state Wildlife and Marine Resources Department. Knox found short hairs and additional tracks, which made him believe the culprit was a red fox.[5] The Wayes and the media were unimpressed.

These events became public knowledge by July 19, 1988, and a monster hunt ensued. By July 21, the newspapers reported that seventy-five hunters were prowling Scape Ore Swamp's edge looking for Lizardman. Sheriff Truesdale said, "Last night it looked like a football game out there. They had a fire going and [set up] campers."

Wildlife authorities expressed concern, but only for the local alligators, which, at the time, were protected as an endangered species. John Evans, a state spokesman, said, "If they kill alligators, we will be on them like a chicken on a June bug."[6]

Lee County Sheriff Liston Truesdale remained dubious as he tried maintaining order in an increasingly high-profile news story. "Something like this just happens. People's imaginations are running away with themselves. If anything, we want to hush things up. It's a lot of extra work for us."[7]

Adding to the excitement, radio station WCOS in Columbia offered a reward of one million dollars for the capture of Lizardman.[8] Monster t-shirts and caps went on sale in Bishopville. And Chris Davis began giving interviews for a fee.[9]

Bloodhounds were brought in from the State Law Enforcement Division (SLED). They might have been defeated by a rainstorm that wiped out the creature's scent. The searchers did find tracks, which SLED Hugh Nunn described: "They had three toes and dug down into the ground in six-foot strides . . . I can't even say this with a straight face."[10]

On July 24, a state trooper and a Lee County deputy investigated a report by four teenagers that something tall with glowing eyes crossed the road in front of them southwest of Bishopville. Two of the teens were Rodney Nolf and Shane Stokes from the community of Turkey Creek. They were out "cruising" with two girlfriends when they all saw something cross the road twenty feet away from them. Stokes told the *Houston Chronicle*, "It was too big to be a deer. It stood on its hind legs, but was too muscular to be a man, seemed to be about seven feet tall. It turned its head once and its eyes glowed, don't know exactly what color. But it ran across the road, past the bridge and into the swamp where Interstate 20 meets Highway 15."[11] The two boys wanted to investigate themselves, but the girls wouldn't have it. So they reported the sighting to the authorities.

The two lawmen who followed up on this report were State Trooper Mike Hodge and County Deputy Wayne Atkinson. They checked out the site and then patrolled near the swamp. Venturing down a tree-shrouded dirt road leading into the swamp, both saw three forty-gallon drums that had been dragged from a dump, crushed, and scattered onto the road. A few saplings had their tops ripped off about eight feet up. Then they noticed indentations in the hard red sand.

"They were humongous footprints," said Hodge.

Each track stretched fourteen-inches long and seven inches across, making deep impressions into the ground. The officers followed the tracks for about four hundred yards.

They drove to the end of the road, returning in about five minutes. "Whatever it was had already walked back across right where we had been. There were fresh new tracks across our tire marks," said Hodge. "Both of us had a sixth sense that somebody had been out there watching us."

Atkinson believed the teenagers: "I didn't doubt they saw something. They were visibly upset."

When Atkinson found the track, he was also impressed by them and the stride, "What worries me is that I'm 6-foot, 3½ inches and I couldn't reach that far. I also tried stomping my foot in the ground that day and couldn't make a mark. Something heavy did that."[12]

Three casts were made. Then a two-inch rain wiped out the tracks before more of a record could be made. This discovery of prints by law enforcement gave the whole series of reports greater credibility. The imprints were large and unusual. The tracks showed three long toes and a high instep. The Associated Press circulated a photograph of sheriff deputies Jim

Tallon and Phil Stokes holding up two enormous plaster of Paris blocks showing the prints.[13] Newspapers around the country featured the story. Another widely circulated photo was of Davis holding one of the casts of this peculiar imprint.

As the summer passed, spotty reports of what could have been Lizardman were made. "Lizard-like footprints" were spotted on a horse trail in the Sumter National Forest, one hundred and fifty miles away.[14] Something was seen in a field near Elliott, forty miles to the west. The last publicized sighting was something seen crossing a road near the swamp on August 26. An Army colonel made the final report to Sheriff Truesdale. The information was remarkable, for the colonel described something different. He said he saw a reptile-like creature walking on two legs with a long tail.[15]

When reports of strange things get a lot of publicity, people tend to speak up about topics they otherwise would keep silent about. Perhaps there are more unknown things on the loose in the South Carolina swamps than just Lizardmen.

For example, the wetlands around the estuary of the St. John River in Florida have a peculiar history. For many years, there have been periodic reports of a pink reptilian mystery resembling *Thescelosaurus*, a Late Cretaceous dinosaur.[16] Elsewhere in the South, people began speaking up about seeing strange tracks and even Lizardmen. Someone living outside of Charlotte said he had seen them for years, but he wanted his name and location kept out of the news. He was successful.

Dean Poucher in Beaufort County, South Carolina, said he had seen tracks seventeen inches long on Old Island in 1970. He added that people had seen tracks and heard sounds in the marshes for years.[17]

As far away as Pass Christian, Mississippi, ten-year old Jeremy DeCoito remembered his experience of seeing the very same kind of Lizardman when he was five. He had been staying with his grandparents at the time. He was examining the pictures on his grandmother's wall when he turned to see what was tapping at a window. He recalled it as having red eyes, green scales, and big teeth. The two looked at each other for a minute. Then the Lizardman ran around a corner of the house and shook the back door. Jeremy's grandmother, Dee King, recalled to Peterson how upset the boy was at the time, screaming and jumping into bed with his grandparents. He spoke up again in 1988 after he saw the drawing done by a *Houston Chronicle* artist. When he saw the sketch, he responded, "That's him!"[18]

In South Carolina, a legend of the swamps had been uncorked. Like the modern birth of Bigfoot in 1958, there were large and peculiar footprints on display. Like the origins of the California Bigfoot, those hefty casts of imprints raised lasting questions about what could make such dents in the ground.

The feet of the water apemen are spectacular in every way. Their very impressions on the ground have struck fear into their hearts. Their unusual appearance has caused scientists to dismiss them as unreal. But they have the potential to tell us something about the life of a merman, as animal tracks often do.

Three primary toes are one of the most apparent features of the footprints left by Lizardmen. The track is not the foot, however. The presence of a fourth toe has been detected. It appears in some of the imprints as an impression beside the main body of the foot. The probable nature of this fourth toe will be discussed later. A high arch is characteristic of Lizardmen tracks. The impression narrows in the middle and sometimes does not register on the ground. In such cases, there is a forward impression of a triangular shape with three prominent toes and a rear one of a circular or elongated shape. The imprint is generally wide at the front and back extremities. Impressions of large claws are sometimes evident at the ends of the toes. The track length varies from fifteen inches to almost twenty inches.

These creatures have three large toes because their limbs are weapons. As humans, we are most familiar with the appearances of our feet and hands, a large toe and smaller toes, and a thumb and four fingers. These can be used as weapons. But in the world of nature, they are relatively puny. That is why we have famously developed hand-held weapons. Long ago, we used bones, wood, and stone to fashion combat instruments. We have refined those tools considerably over thousands of years. We have worked our way up from clubs, through crossbows and firearms, to bombs and lasers.

On the other hand, merbeings faced different demands in a watery world. They had to defend themselves with their very limbs. And when the primate limb is employed as a weapon, the power of three digits with claws becomes the best option.

Try extending three of your own fingers, and imagine using them as a weapon. You will sense how insubstantial they appear. Now imagine that at the end of your arm are only three digits, each sporting a sharp claw. Now that would be a weapon!

There is a parallel in the primate world in developing a three-toed limb. However, the owner of those limbs is yet another cryptid. They are called Giant Monkeys. They survive in Africa, Asia, and the New World. Their tracks show three large toes, although they are long and thick.[19] These monkeys are the survivors of *Simopithecus*, the largest monkey ever known. Their tracks also appear as large impressions. It's possible they developed their spectacular feet for the same reasons. The feet were formidable as weapons that could have contributed to the survival of *Simopithecus* into modern times. This contention is distressing to those who also dismiss the presence of mermen and mermaids.

These primates have a five-digit limb in common with other animals. But for them, the limb evolved to a point where three digits were the primary ones. The others either developed other functions (as we see in the case of water apes), or the digits became vestigial and no longer apparent when tracks are left behind.

The association of the three-toed track found at Bishopville with the reported Lizardman of 1988 is a crucial link in our examination of this topic. The tracks at Scape Ore Swamp were detected by law enforcement personnel. They discovered them crossing their own tire tracks that they had just left minutes before. They made casts and displayed them to the public during the excitement. What they found can reasonably be tied to what was being reported at Scape Ore Swamp.

In truth, the 1988 Lizardman encounters may have been the last sightings of a native species, with reports dating back four hundred and fifty years. In 1530, *Orbe Novo*, a history of Spain and its New World discoveries, was published.. It includes an interview with Francisco de Chicora, the baptismal name given to a Native American kidnapped by Spanish slavers in 1521. Chicora's data on the Carolina coast includes the village of Inzignanin, often spelled Inisiguanin.

> There is another country called Inzignanin, whose inhabitants declare that, according to the tradition of their ancestors, there once arrived amongst them men with tails a metre long and as thick as a man's arm. This tail was not movable like those of the quadrupeds, but formed one mass as we see is the case with fish and crocodiles, and was as hard as a bone ... Their fingers were as long as they were broad, and their skin was rough, almost scaly. They ate nothing but raw fish, and when the fish gave out they all perished.[20]

Chicora was abducted at Winyah Bay, South Carolina, claiming this tale was passed down by his local ancestors. This becomes important when it is realized that Winyah Bay is the mouth of the Pee Dee River. Among the Pee Dee River tributaries is a familiar name, Scape Ore Swamp.

ENDNOTES

1 Stein, "Bigfoot believers." *Denver Post*, 05 January 2003.
2 Hall, *Thunderbirds*, 14.
3 "Fame Follows Close Encounter of the Lizard Kind." *Charlotte Observer*, 02 August 1988.
4 Tuten, "First Lizard Man spotter passes polygraph." *The State*, 03 August 1988.
5 "'Lizard Man' sightings rouse some excitement." *Anderson Independent-Mail*, 21 July 1988.
6 Tuten, "Pet alligator not monster, sheriff says." *The State*, 21 July 1988.
7 Georgas, "'Lizard' calls besiege Lee sheriff." *The Item*, 21 July 1988.
8 Tuten, "Monster bash – 'Reputable people' say they saw the elusive 'Lizard Man.'" *The State*, 20 July 1988.
9 "Teen Who Spotted Lizard Man Reaping Benefits." *The Aiken Standard* 02 August 1988.
10 Tuten, "'Skunk ape' still causing a stink." *The State*, 27 July 1988.
11 Horswell, "On a scale of one to 10, it rates a downright scary 11." *Houston Chronicle,* 31 July 1988.
12 Tuten, "Trooper convinced that 'something is out there.'" *The State*, 28 July 1988.
13 Drape, "Lizard Man a Tall Tail?" *The Atlanta Constitution* (Atlanta, GA), 31 July 1988.
14 Squires, "Lizard-on-the-lam feared in area." *The Index-Journal*, 25 July 1988.
15 "The Lizard Man returns; seen by 'respected' citizen." *The Item* (Sumter, SC), 02 September 1988.
16 Hall, "Pinky, the Forgotten Dinosaur." *Wonders* 1(4), December 2002.
17 Tuten, "Tracks pique public interest." *The State,* 15 August 1988.
18 Peterson, "Lizardman has Coast connection." *Sun Herald*, 13 August 1988.
19 Hall, "Guide to North American Monsters," *Wonders* 6(1), March 1999. See also, Coleman and Huyghe, *Field Guide* (2015), 34–37.
20 d'Anghiera, Pietro Martire. *De Orbe Novo* (1912), 261.

Chapter Twelve
Lizardmen of the Carolinas

THE COASTAL REGION OF THE CAROLINAS is a natural home for wildlife oriented to a watery habitat. At the heart of this region is the Carolina Low Country, which extends from the Savannah River to the Green Swamp of North Carolina. Other swampy areas continue up the coastal plain into Virginia. Among them are the Angola Swamp, the bottomlands of the Roanoke River, and the Great Dismal Swamp.

In this world of swamps and riverine forests, merbeings do not appear to be out of place. They have left traces in the geography and history of the Carolinas. Mermaid Point is found in Chatham County of North Carolina. You will find it where the Deep and the Haw Rivers combine to form the Cape Fear River. Malcolm Fowler gave this origin for the name in *The Story of Fayetteville*, published in 1950.

> "The meeting place is called Mermaid's Point. Tradition says that in ages long gone, mermaids, tiring of the tamelessness of the Atlantic, would swim up the Cape Fear, battle their way through Smiley's Falls above Erwin and the Great Falls at Buckhorn to relax on the white sands of the Point. There they would sing mermaidish songs while they washed the sea salt from their tresses. Many travelers, weaving their way homeward at night, swear they have heard the murmuring sounds of the mermaids' singing."[1]

One of these merfolk seems to have turned up in a fisherman's net in North Carolina's coastal waters. However, it was not recognized as such and became an odd tale and the subject of speculation. The account was preserved by a man named J. N. Blevins in *The State Magazine* in 1967. He was reminiscing with people in northeastern North Carolina about things that had happened from Dismal Swamp down to Hyde County when he heard this story:

> All of them had heard about John Patrick, and the man-like creature that he caught in his fish net one night at about the point where the Pamlico Sound backs up into Long Shoal Bay. They all agreed that John Patrick was known to be a truthful and honest man, and that he started to pull up his net and something crawled up on the side of his boat, and he beat it off of the boat with a paddle, and never went fishing again in Long Shoal Bay. They all agreed that it was supposed to be a man-like creature.[2]

Blevins offered his own solution to this incident. He thought it might have been a frog man from a Russian submarine. Such speculations are not unusual when people are trying to explain away mysteries. Another ready suggestion has been that such things come from outer space by way of Unidentified Flying Objects. Some of this was also done during the Lizardman excitement of 1988.[3]

As mentioned previously, in 1988, Chris Davis referred to the Scape Ore Swamp creature as "a red-eyed devil." This brings to mind the historical record from two centuries ago in North Carolina. Frequent sightings of "the Devil" were reported in the 1795 autobiography of the Rev. William Glendinning. Starting in 1785, Glendinning claimed to see the Devil when he visited and preached in Granville and Halifax Counties of North Carolina.

The silent appearances of the "devil" seen by Glendinning could be encounters with a merbeing. The one exception, where a voice is attributed to the creature when it was standing outside a window, might be a misremembered detail by the time Glendinning wrote his memoirs. A silent encounter is similar to the accounts in 1988. Here is what Glendinning experienced, beginning when he opened his door in response to a loud rap and came to face to face with the thing for the first time:

> Immediately there was a loud rap on the door, I opened it, and saw his face: it was black as coal—his eyes

and mouth as red as blood, and long white teeth gnashing together. I shut the door, and in a little time he went off: and some present then observed that they never heard it blow or rain harder.[4]

As the winter continued, Lucifer would appear two or three times a week, either in the evening or before the family went to bed. Then one day, Glendinning was a short distance from the house where he was staying...

> When Lucifer appeared in view at a distance from us. He made toward us, then turned toward the spring branch and disappeared. It was about noon when he appeared, and the first time I ever saw him in day-light. He appeared upward of five feet high—round the top of his head there seemed a ridge;—some distance under the top of his head there seemed a bulk, like a body, but bigger than any person—about 15 or 18 inches from the ground there appeared something like legs, and under them, feet; but no arms or thighs. The whole as black as any coal; only his mouth and eyes as red as blood. When he moved, it was like an armful of chains rattling together. Several of the family were present—they declared they saw it, and said, they never beheld the like in their life.
>
> Frequently during the summer, in the year 1786, and for near two years after, he would appear.—For the most part, he made his appearance in the fore-part of the day. He would often stand in the orchard, near my cabin, and shoot out of his head something like a horn, about six or eight inches high, above the top of his head.[5]

The curious features of a ridge on the top of the head and the appearance of lacking arms (but not necessarily lacking them) are traits often attributed to merfolk. This historical curiosity might be one effort to categorize the otherwise inexplicable appearances of a water ape to people who were not prepared for them.

A record of merbeing sightings appeared in small newspapers in late June of 1892. The account, attributed to the *St. Louis Republic*, sporadically appeared nationally in newspapers for fourteen months.

The People residing along the Palmetto creek, South Carolina, as well as those for miles back in the "slashes" are highly excited over the appearance of a strange and uncouth creature in that vicinity. The beast is described as being a creature that far outdoes the nightmares of mythologists. It is equally at home in the water, on the land, or among the tall trees of the neighborhood, where it has been most frequently seen. The general contour of the head reminds one of some gigantic serpent with this exception: The "snout" terminates in a bulbous, monkey-faced knot, which much resembles the physiognomy of some gigantic ape. From the neck down, with the exception of some fin-shaped flippers, which extend from the arms to the waist, the creature resembles a man, only that the toes and fingers are armed with claws two to six inches long.

Tracks made by the beast in the soft mud around Hennis Lake have been taken to Donner's Grove, where they are kept on exhibition in a druggist's showcase. Those who have seen the horrid thing face to face say that it is a full nine feet in height, which could hardly be believed only for the fact that the tracks mentioned above are within a small fraction of fifteen inches in length. Fishermen who surprised the monster sitting silently on a mass of driftwood declared the back looked like an alligator's, and that it had a caudal termination a yard long, which forked like the tail of a fish. It has been seen a number of times sporting in the deepest water in Hennis Lake, then again while running at greased lightning speed in a dusty road, and finally while swinging from tree to tree with as much ease as the most active monkey."[6]

The place names do not exist in South Carolina. What is most uncanny are the parallels between this potentially fictional sighting in 1892 and those reported in 1988, suggesting the possibility that the encounter did occur but with identifiable details altered to save the original witnesses from ridicule. The fascination with the tracks and the appearance of the creature with its claws are certainly familiar. The forked tail has more than one possible explanation. Perhaps the observations were confused, similar to the 1988 report by the Army colonel of a long-tailed creature.

However, some sketches of merbeings originating in North America and Europe show such an appendage on a creature with two legs. In that case, we can see two other possible explanations. One theory would be a naturally occurring tail on some of these creatures. Some animals, like kangaroos, have utilized their long tails as a balance which allows them to remain bipedal on uneven terrain. The alternative would be that this forked tail appendage is some accessory or costume.

Another account almost a hundred years old has been found in Virginia. It has many puzzling aspects. If it has any basis in fact, it offers some observations of a female Lizardman. According to a mountaineer named Stone Colby, who led an expedition from Jamestown to investigate a weird tale regarding a snake woman who is said to frequent the wild parts of moonshine country:

> The snake woman is about 35 years of age and in physical conformation and habits so closely approaches the reptile species that she might be regarded as a missing link between it and the human race. Stone declares that she is more like a snake than like a woman. He has never seen her assume an erect position. Covered with the scaly skin of a snake and shedding it twice a year in one place the snake woman glides among the trees and rocks in search of the small animals, mice, frogs, ground squirrels and other forest and swamp prey, which Stone has seen her eat alive, swallowing them like a reptile, without mastication. [7]

The "snake woman" appears similar to the merbeings seen in 1988, while the crawling behavior might be used to surprise and overcome prey. Depending on terrain, it may also show that merbeings switch between bipedal and quadrupedal. Further delving into early newspaper files gives us more information about these and similar events.

An indication of the long-time presence of merbeings in the Carolinas can be found in the curious reports of what we will, for our convenience, label as "mounting behavior." In the famous 1988 episode, the creature leaped upon the automobile of Chris Davis. Davis is not alone in reporting such an occurrence.

In fact, reports of merbeings joining a ride behind a terrified traveler go back for hundreds of years, and these reports are international. An integral part of Japan's lore around the diminutive aquatic *kappa* involves their attempts to lure horses into the water.[8] Is this mounting behavior culturally related to the *kappa* by a common ancestor?

There is a place in Lincolnshire, England, called Bayard's Leap. This local story tells of a champion that appeared to free the land from the tyranny of a witch but first had to obtain a horse. He selected Bayard from the village herd based on how the horse reacted to a stone tossed in the water. Riding Bayard, he challenged the witch. In the battle, the witch leaped upon the horse behind the rider. The frightened horse took three giant leaps and unseated the enemy, enabling the hero to slay her.

The folk figure of the horse Bayard dates back to *Les Quatre Fils Aymon* in late twelfth century France. It was so popular that it spread across Europe. In this later English tale, the focus for us should not be on the historical overlay but on the details, which should appear familiar: a traveler wandering near water is ambushed by an unknown creature of exceptional speed and strength. As one local historian notes, the earlier versions of the witch had "such terrible claws it may be gathered that she was scarcely human in form."[9]

Another record is specifically dated and placed precisely in nearby Staffordshire. Notice was set down in a volume of Shropshire Folklore as a "ghost animal" account. The rider could only identify his assailant as "a strange black creature with great white eyes." The rider was approaching a canal west of Stafford. This same canal remains on British maps as the Shropshire Union Canal. It has been the site of repeated mounting attacks. Here is one such account.

> On the 21st of January, 1879, just before [a laboring man] reached the canal bridge, a strange black creature with great white eyes sprang out of the plantation by the road-side and alighted on his horse's back. He tried to push it off with his whip, but, to his horror, the whip went *through* the Thing and he dropped it on the ground in fright. The poor tired horse broke into a canter, and rushed onwards at full speed with the ghost still clinging to its back. How the creature at length vanished the man hardly knew...
>
> Now comes the curious part of the story. The adventure, as was natural, was much talked of in the neighbourhood, and of course with all sorts of variations. Some days later the man's master (Mr. B----- of L-----d) was

surprised by a visit from a policeman, who came to request him to give information of his having been stopped and robbed on the Big Bridge on the night of the 21st January! Mr. B————, much amused, denied having been robbed, either on the canal bridge or anywhere else, and told the policeman the story just related. 'Oh, was that all, sir?' said the disappointed policeman. 'Oh, I know what *that* was. That was the Man-Monkey, sir, as *does* come again at that bridge ever since the man was drowned in the Cut!'[10]

Another story reported from the Georgia side of the border with North Carolina also suggested a supernatural origin.

"In Towns County, while a farmer was returning home late at night, a white object, in the shape of a man, darted from the bushes and kept running ahead of him. The farmer spurred his horse and endeavored to overtake the fleeing figure; but it still kept ahead of him for the space of at least a mile when it vanished in the darkness. The negroes in that section are considerably excited over the story, and declare it is "a hant" which bodes no good for their race."[11]

In the early 1900s in North Carolina, we find a history of mounting attacks near Avery County's Slippery Hill Cemetery.

Young men who rode horses by Slippery Hill Cemetery in the early 1900s told of feeling a presence jump on their horses behind them as they rounded the first curve at the graveyard.

Uncle Seth Waycaster, who died in 1954, told of riding his high strung gelding by the cemetery on a cold March night around 12 o'clock.

"I was riding hard any way," Uncle Seth said, "because I had heard about the [ghost] lights and didn't want to see one. All at once I felt something land right behind me on the horse's back, and wrap long cold arms, like steel bands, around my body."

"I tried to look back over my shoulder but could not, but I could feel its icy breath on the back of my neck. My hair felt like it was standing on end, and my horse was terrified; it bolted and reared, like to have 'a throwed me. I hung onto the reins for dear life and when I reached the end of the cemetery, the being vanished."

"I never rode by that graveyard again at night, but took the other way home."[12]

The author, Bertie Cantrell, also noted two recent vehicular accidents at this same location within two years:

> The most recent victim had lost his life when his car careened out of control at the end of the cemetery, crossed the road and went 200 feet down the steep embankment leading to Toe River.
> The young man who died almost two summers ago had been on a motorcycle that left the road at the south end of the graveyard, crossed the road and plunged over the embankment. His body was not discovered for a week.[13]

Cantrell closes the article with a reasonable comment: "Those of us familiar with the legend let the investigators put cause of the recent accidents down to excessive speed and a bad road, but we wonder if the Phantom still lurks on Slippery Hill Graveyard, leaping onto the vehicles as he did onto horseback in other days."[14]

Mountings, in the horseback days, were widely reported. In Ashe County, odd things were associated with a spot known as the Devil's Stairs. Robert E. McNeill sought out folklore in the county. He recorded an account told by Harry Dollar of what happened to his grandfather in the 1910s.

> My grandfather, W.T. Dollar, he was a drummer—a horseback drummer for R.J. Reynolds Tobacco Company of Winston-Salem, and he drummed tobacco, you know, sold snuff and chewing tobacco. And the first thing, he always rode up by White Top through Konnarock when he was heading home, and he would always have to ride by the Devil's Stairs. Usually, dark would catch him, and one night—I heard my father tell this now—one night, my grandfather came by the Devil's Stairs at sundown,

and something got on the horse behind him, you know, and his horse started fretting, and he couldn't control it. His horse got kind of out of control with him, and which there's a church still standing, known as the Oak Grove Baptist Church, now. When he come by there—when he got to the church—the horse just came to a complete stop without his controlling it, and when it did, why, it was just something getting off his horse, and he rode on in up, which he lived about, oh, three-quarters of a mile up the road at the homeplace...

He said when he got home, there was something, something on his horse. It looked like white marks of some kind, something you might say looked like chalk marks or something in the shape of somebody's legs on his horse's flanks, you know. He wiped to see if it was sweat from where his horse was a-fretting, but it wasn't, he said; it wasn't sweat. It was white marks like somebody had gotten on there with something on behind the saddle there. Whatever it was, wasn't on the saddle there; it was on behind the saddle. Ever what it was sitting there behind him had hung down and touched the horse's flanks, and my grandfather at that time, as anyone could witness, kept the best horses, 'cause he depended on his horse for his livelihood. Now, from that day on, it was always told to me, you could never get my grandfather to pass the Devil's Stairs at night...[15]

From Cleveland County, an account of a "headless woman" leaping on a horse was taken recorded in 1927:

John Gantt, of upper Cleveland, was riding along a country road one night with a friend and at a certain place in the road, a headless woman appeared. She followed a short distance, and jumped on the horse behind his friend. She remained there only a few moments until she vanished. Mr. Gantt resolved to ask who she was and what she wanted if he saw her again. So, he happened to be passing by there a few nights later, and she came and jumped on the horse behind him. He didn't ask her anything.[16]

More reports of repeated mounting attacks come from the south-eastern corner of Cleveland County, ranging from horseback days up to modern transport.

Back during the horse-and-buggy days he'd spot a lone rider, materialize from out of the woods, and leap on the horse behind the rider. He'd cling to his terrified bene-factor so tightly the poor fellow couldn't jump or stop, only spur his steed ahead at its quickest pace. When his destina-tion was reached, the ghost would alight and his shapeless form would be seen streaking off into the woods again.

But in 1950 he selected a more modern conveyance. A group of Kings Mountain hunters had been stalking rabbits on Milk Dairy Creek in the Linwood area all after-noon. No one seemed to notice when darkness settled in. Their dogs were still jumping and running rabbits and the hunters were as enthusiastic as when they started.

Suddenly "something" white dashed from the swamps and headed their way. The men took to their automo-bile (a 1939 model with running board) with lightning speed, leaving the whining, tail-dragging dogs to fend for themselves.

The driver jerked the car into gear and started down the old sawmill road. The ghost jumped on the right fend-er, pressing his weight forward till the vehicle took a de-cided list to starboard.

Let your window down and I'll empty my shotgun into him," a backseat passenger shouted. But the rider up front was frozen with fear. He didn't move. Just before the careening car reached the highway the ghost jumped from the fender and headed for a plum thicket.

The driver kept his foot to the floorboard. Nobody looked back. The passengers just kept their eyes straight ahead. They weren't waiting for any thanks for the lift. They didn't even take time to figure why a ghost had to ride anyway. Why couldn't he walk—or float.[17]

In 1975 Douglas Fisher interviewed Gordon H. Cameron, a resi-dent of Elizabeth City. Elizabeth City is just south of the Great Dismal Swamp in the eastern part of the state. About eight miles outside of town on Highway 17 is a stretch of road called the "Four-mile Desert." Only scrub brush grew there. A bridge at one end was said to be haunted. Around 1966–1968, Cameron had driven out to the haunted bridge because of the legend. The local story was "a young decapitated woman was discovered beside the bridge some years back, with rather unusual claw-marks covering her body." No one was sure who or what was haunting the bridge. Theories ranged from a monster to some sort of wild man. Cameron offered Fisher a first-hand account of what happened on his next visit to the bridge.

You know, Kenny and myself were spending quite a bit of time together back in those days ('66, '67, and '68). We had heard a lot of stories about the four-mile desert for years, and had even been by there several times during the daytime. Anyway, we had been seeing these two girls from the high school, and one night, having nothing else to do, we all decided to grab a couple of bottles of wine and make a run out to the four-mile desert.

Kenny didn't have any wheels at the time, and neither did I, so we had to take his mother's car, a new Ford station wagon. We were a little worried about that at first, since the road was kind of rough, and it had just rained a couple of days before to boot. But after we got the girls sort of scared telling them all the stories we had heard about the place, we were even more determined to go, so we soon quit worrying about the car.

After we got the wine, we rode on out, me and Kenny feeling fine, and getting those chicks more scared all the time telling them the stories we had heard about the place. Now you've got to understand that that place is really eerie, no matter how many times you've seen it before. As we got close to the bridge, even Kenny and me were beginning to feel a little nervous, even though we had never taken those stories very seriously. Anyway, we pulled over to the side about twenty or twenty-five feet this side of the bridge and rolled down the windows just a little bit. The girls were really uptight by that point, because the bridge over that little creek was a freaky-looking thing, and the creek itself was real weedy and swampy, and combined with the noises from the frogs and crickets and things, it was enough to make almost anybody a little bit ill at ease.[18]

The courage of the two young men returned after emptying the wine bottle. They exited the vehicle and sat on the hood, while the girls remained inside. The windows were rolled up, and the doors were locked. Next, they all heard a splashing in the water beside the bridge. Something then began to climb up the bank twenty feet from them.

> We had just scrambled into the car when we saw a dark shape coming toward the car from our right. Kenny ground the starter, and for a few seconds the car wouldn't start. That really freaked us out, because one of the things we had heard about before was that various people had disappeared at the bridge because, once they turned their car off, it wouldn't start again. The damn car finally started, and Kenny threw it in drive and put it to the floor.
>
> We just missed whatever had come up from the creek, and all of a sudden we felt the rear of the car weigh down. The girls started screaming and buried their heads. Kenny started swerving from side to side and, as soon as the road was wide enough, he slowed down, skidded the car around, and headed back for Elizabeth City. No matter how he drove or how fast, the rear end of the car was still weighed down.
>
> Anyway, no matter what Ken did, we couldn't shake the thing until we were almost back to Highway 17. Finally, about a couple of hundred yards from the stop sign, we felt the rear end of the car lighten up. I did muster up the guts to look back then, but I couldn't see anything at all. Kenny was sweating and the girls were almost hysterical. I was too freaked out to do anything except keep yelling, "What the hell was that?"[19]

By the time they returned to Elizabeth City, they had calmed down. The car needed gas, so they stopped at a station on Main Street. Then, as an attendant filled the tank, he called their attention to the car's roof.

> On top of the car there were two sets of deep scratches, about two or three feet apart. There were five scratches

on each side, and they were a couple of feet long. When we saw those scratches, we really got so shook that Ken took us all back to his house.

Ken's mother didn't believe a word we told her. I guess you already saw that coming. Anyway, he had to pay to get the top of the car fixed. We talked about what happened that night for a pretty long time, but after awhile we just sort of lost interest. Anything gets old after a while.[20]

These incidents appear to mark the behavior of similar creatures. They have been playing this trick for over a hundred years, first by mounting behind riders, then adapted to jumping on top of vehicles when they replaced the horse. The incidents occurred at night and were over so fast that people did not get a good look at what was creating the disturbance. Instead, people have turned to ghosts for an answer, as they often do in search of a solution to a mystery.

The act of leaping upon horses and vehicles is such a trivial act that it is surprising that we have so many cases on record. It is the kind of experience for people that might go unreported in many cases if the Lizard-men were doing this routinely. There is no obvious context for the human participant. It is not a story where the teller gets to be the hero. He is, in all ways, a victim. When someone tells this story, they are putting themselves in the role of a person startled, frightened, helpless, and having no explanation for the event. He is more likely to be disbelieved than to be accepted. Therefore, one is discouraged from speaking up.

These stories have found their way into the chronicles of folklore as ghost stories. This comes about from the need of folklorists to find some context and create associations when compiling their data. This activity has even been given its own designation under the heading of "The Dead." In the past, when merbeings attacked horses, it was more acceptable to claim an encounter with demons, the Devil, or ghosts. Somehow, this made more sense to victims than admitting that they saw creatures made of flesh and blood, not dissimilar to ourselves.

The presence of stories both in the United Kingdom and in the Carolinas calls for further comment. Folklorists with a traditional mindset will take this as a sign that the entire business is explicable as a twice-told tale. For them, it will be a fictional story related repeatedly. It would travel from continent to continent in the minds of immigrants who will have transferred the locale for the sake of compelling storytelling. Only in such a manner will this folklore make sense to them.

The unique behavior of water apemen "mounting" horses and vehicles must have some meaning that remains mysterious to us. A couple of possible motives come to mind. One is that this action is important to the water apemen among their own kind. For example, it could be a rite of passage, akin to killing a lion to mark an entry into a new tier of their culture. Perhaps the act bestows the status of seeking out an encounter with human beings and getting the better of them.

Another possibility is that they simply do it for fun. It might be a good story among themselves, the Lizardmen's equivalent of a centuries-long running joke about how easily they can spook the hapless and unsuspecting humans. And perhaps, they still are playing games. In September of 2021, Ruth Teasley, a Savannah, Georgia, resident, awoke to find her car damaged in a manner that brings to mind the damaged vehicles of the Wayes in South Carolina back in 1988. Both of her wheel wells were ripped off. Wires were chewed off, and teeth marks, scratches, and paw prints were scattered across the car's front. Teasley claimed she heard a stray dog barking the night before and blamed the stray. The County Animal Services Director had never heard of a dog attacking a car and remained unconvinced.[21] At least this time, it was blamed on something larger than a fox. It should be noted Teasley lived less than two thousand five-hundred feet from the Springfield Canal, which empties into the Savannah River less than three miles away.

For me, this preliminary suggestion of widespread occurrence opens up the consideration that this behavior will be found to be something deeply rooted in the lives of merbeings. It also raises the question about accounts of attempting to climb into boats, as discussed in Chapter Three and Chapter Four. Are boat boardings a variation of the land-based mountings? Only further research will establish that such events are genuinely widely recorded. I will not be surprised if they continue to turn up.

The old stories of merfolk following ships might have some basis in fact. The creatures could have followed ships and brought their ways with them to a New World, just as human immigrants have done. So possibly, we are seeing the result of such a migration. This peculiar interaction with people seems to be occurring globally.

ENDNOTES

1 Fowler, "A Tour of Cape Fear," in *The Story of Fayetteville and the Upper Cape Fear* by Oates, 337.

2 Blevins, "Stories the Goose Hunters Never Hear." *The State* 35(10), October 1967. 19, 32.

3 Tuten, "Lizardless Summer," *The State*, 24 July 1989.

4 Glendinning, *The Life of William Glendinning, Preacher of the Gospel* (1795), 19–20.

5 Glendenning, 22–23.

6 "An Uncanny Monster." *The Larned Eagle-Optic*, 24 June 1892.

7 "Hunt Animal Half Snake Half Woman." *Philadelphia Inquirer*, 21 August 1907.

8 Eiichirô, "The Kappa Legend," *Folklore Studies* 9 (1950).

9 "Bayard's Leap" *Lincolnshire Notes & Queries* (1902–03), 211–214.

10 *Shropshire Folk-Lore; A Sheaf of Gleanings* (1883), 106–107.

11 "Queer Things in Georgia." *Atlanta Constitution*, 20 April 1895.

12 Cantrell, "The Legend of Slippery Hill," *The State* 49(1), April 1982. 15, 31.

13 Cantrell, 15.

14 Cantrell, 31.

15 McNeill, "Legends from the Devil's Stairs," *North Carolina Folklore Journal* 26(3), November 1978. 152.

16 Brown, "The Headless Woman" *Frank C. Brown Collection of North Carolina Folklore,* vol.1 (1952), 680–1.

17 Smith, "The Ghost that Rides," *North Carolina Folklore Journal* (May 1975), 44–45.

18 Fisher, "The Four-Mile Desert: A Horror Story," *North Carolina Folklore Journal* 23(2), February 1975. 22–23.

19 Fisher, 24–25.

20 Fisher, 25.

21 Bolden, "Savannah woman wakes up to car severely damaged by unknown animal." *WTOC*, September 10, 2021.

Afterword: What Can We Learn?

IF PUBLISHED ACCOUNTS OF SIRENS were enough to establish the existence of the creatures, they would have entered our nature handbooks long ago. They have remained enigmatic because we will do nothing about them until we have a dead one to dissect.

I do not want to dissect a mermaid to prove they exist any more than I want to dissect a person to establish that people inhabit other lands. We no longer approach people that way, though such an approach was entertained five hundred years ago. I find the merfolk not enigmatic, although they are very intriguing. The merfolk exhibit a variety of appearances. This indicates that they have adapted to varying water environments, and that these adaptations have sustained them over the centuries.

They appear to be intelligent beings who are worth our attention and appreciation. We must put aside any preconceived notions and examine them with an open mind and genuine fascination.

Nine million years ago, the primate *Oreopithecus* spread around the world into most aquatic environments. The creatures' diversity of appearance is due to natural selection. These variations occurred for survival, and reflect their need to adapt to different circumstances, including marine and freshwater environments, differing food sources, and predators. Buffeted by natural geological upheavals, the different-looking merbeings have returned from such setbacks to find ecological niches and persevere.

We know a lot more about the parallel activities of land primates. Over a similar stretch of years, they proliferated in lines of evolution, giving us many competitors with similar characteristics. The results of this are *Paranthropus* (known popularly as Bigfoot), the Little People (*Homo floresiensis*), *Homo erectus*, Neandertal men, *Homo gardarensis*, and the True Giants (*Gigantopithecus*). With such variety in land primates, a parallel in the water-portion of the planet, which is much larger, appears to be a natural occurrence.

Today the pressure of competition from humans also extends to their aquatic cousins. The world of freshwater habitats is reduced constantly. Marine habitats are being affected by pollution. The shore areas of the landmasses that were once rarely visited are now becoming more developed and populated. Merpeople have been forced to adapt to this.

In ancient historical records, there are indications that merbeings willingly and regularly interacted with human beings. But over time, the interactions decreased, and they disappeared from the awareness of people. They became legends.

Despite the evidence for the existence of merbeings, we have remained largely ignorant about them. It could be argued that our collective arrogance regarding them is second nature. We have had great success in pushing aside relatives like *Homo floresiensis* and *Homo gardarensis*, to mention a couple whose physical remains were discovered. Because we are in possession of physical evidence, they have been assigned scientific names. The living aquatic survivors of these evolutionary lines appear not to matter to us in the modern world. Because we don't have physical evidence of their existence, some might assume they have died out, or aren't worth searching for because they didn't evolve to take over on land, as *Homo sapiens* have. They could be seen as "losers" in the scheme of evolution, since they're relegated to a life in the water.

There are a couple of good reasons not to look at merbeings this way. They are not losers. They are remarkable survivors. Humans have been rising from a primitive ape ancestor and competing for preeminence on the land. The water apes have spent those same millennia doing the same thing under water. They have figured out how to survive in an environment hostile to primates. They have overcome the effects of periodic upheavals on a planetary scale. They have shown that they are not going away any time soon.

They are aware of us and stay away. This is because humans have given no signs that we are willing to accept them or share the planet with them peacefully. The case of the pygmy water apes communicating with the Native Americans in Oklahoma shows us that they are managing their relations with humans carefully. They have the upper hand in knowledge about us, while we are only beginning to piece together some understanding of them.

Furthermore, if we were overcome prejudice and connect with our aquatic relatives peacefully, we could open up a revealing dialogue with these intelligent creatures. We must find the courage to attempt such a dialogue and hope they can be convinced to engage in this exchange. It could be a turning point for humankind. It could change our understanding of the oceans, lakes, rivers, and wetlands..

So, what could we do to learn more about the water apes?

We would benefit from a global review of the literature. This would consist of folklore, detailed reports of sightings and encounters, and various reviews and opinions on the subject. A good example of such literature is John E. Roth's *American Elves: An Encyclopedia of Little People from the Lore of 380 Ethnic Groups of the Western Hemisphere.* His work covers over three thousand "little people" in three hundred and eighty cultures in just one hemisphere.[1] A thorough search for the lore of merbeings will likely result in a similar tome.

In order to compile this, we need to encourage people to speak up about what they know about water apes. The more accounts we have, the more we'll know. But, of course, there should be an established way for people to provide reports, and at this time, there is none.

What is truly needed is a change in attitude in academia. Academics should be at the forefront of gathering new knowledge. Unfortunately, many do nothing to find new information on water apes.

But efforts have been made to get information into academic hands. George Eberhart, senior editor of *American Libraries* magazine for the American Library Association, assembled citations on water apes in his book *Monsters: A Guide to Information on Unaccounted for Creatures.*[2] So there is a way to bring this information to the attention of academia, as the subject of Sasquatch has been brought to academia. The previously-ignored testimony of Native Americans began to be organized by John W. Burns, a Native agent on the Chehalis reservation in the lower Fraser Valley of British Columbia. Starting with a 1929 magazine article, Burns ceaselessly supported the Native stance on Sasquatch's existence in the popular press.[3] Nearly a century later, academics such as archaeologist and anthropologist Kathy Moskowitz Strain have discovered and given credence to the topic.[4]

If academics could be more forward-thinking, it might not take a lifetime to make discoveries of some importance.

In the spirit of academic freedom and curiosity, it would serve universities to seek out and preserve records on cryptozoological topics. These efforts might advance the pace of learning and help reveal cases of neglected wildlife. The resources of the International Cryptozoology Museum (ICM) are an ideal example of how to keep such records. Unfortunately,

ICM is a small, nonprofit organization taking on a herculean task. Imagine if they were part of a more extensive network of universities and museums with the resources, funding, and personnel to preserve information.

So why don't they have access to this network? Because academic institutions lack the will and the courage to pursue cryptozoology. They fear taking on such bold initiatives. The people who work in academia are typically the ones who have been vocal about how distasteful the topics are. Though they do not have the perspective to realize it, they are actually just avoiding the challenging topic of undiscovered animals.

As we have discussed, the reality of merbeings carries a lot of baggage. Like other animals, they aren't easily seen or captured. We want to put things into zoos, and these water apes aren't allowing it. They inhabit oceans, high mountains, swamps, and dense forests. They are too big, fast, intelligent, or rare to fall into our nets.

I will say that it is entirely possible that merbeings are interested in a partnership with human beings. However, I admit I may be anthropomorphizing the creatures. Some will object to ascribing human thinking, feelings, and desires to another distant primate species. To this objection, I will say that some anthropomorphism may be justified. After all, they are related to humans and have been evolving for millions of years.

They have had to fight for their survival against the creatures of the water world. They have invented the means to be active in that environment. They have recovered from geological catastrophes. They have, over time, multiplied and inhabited the globe in parallel to how we have spread as human beings. They have families and raise their offspring, as near as we can tell from anecdotal accounts. Likewise, we know of their liaisons with human beings where offspring have been produced and raised. Could they not have developed similar senses to our own? They might be more like us than we would think. We should look for their similarities, not just the noticeable differences.

If we continue to research and share knowledge of these creatures, over time people will become more comfortable with the idea that they are distant relatives with remarkable abilities. And one day, in the foreseeable future, we will embrace the realization that merbeings do indeed share planet Earth with us.

ENDNOTES

1 Roth, *American Elves* (1997).

2 Eberhart, *Monsters: A Guide to Information on Unaccounted for Creatures* (1983)

3 Burns, J. W. "Introducing B. C.'s Hairy Giants," *Macleans*, 1 April 1929.

4 Strain, *Giants, Cannibals & Monsters* (2008).

Appendix 1
In Memory of Mark A. Hall

Loren Coleman

I WANT TO INTRODUCE YOU TO THE GENIUS behind the core research that resulted in this book. My close friend, colleague, co-author, and fellow Midwesterner, Mark A. Hall, died on the morning of Wednesday, September 28, 2016. He was a profound thinker who helped lay the foundation for cryptozoology. It will be years before young researchers realize this loss. But for me, his death was an immediate, powerful punch to the gut.

Mark's brother had passed away a few years before. His brother's widow, Shelia Hall, became a lifeline for Mark in his last years. Mark had been challenged by cancer for years, and it was a constant fight. Shelia was able to be with Mark the night before he passed away. He talked for over an hour to her, saying again that he was happy his research was in good hands with the International Cryptozoology Museum (ICM). She said that Mark died peacefully.

Mark was born on June 14, 1946, in Minneapolis, Minnesota. It seemed fitting that he entered the world on Flag Day, as he was interested in politics and how the nation should be run. Fortean, cryptozoologist, author, and theorist, Mark was raised in the heartland of America, in or near Bloomington, Minnesota (except for one temporary trial attempt at living in North Carolina). In the midst of the Cold War, Mark served in Army Intelligence in West Berlin as a Russian linguist, translating messages from behind the Berlin Wall. After his military service, he became an editor at an archeological society in Minnesota. He then worked in human relations in various branches of the federal government, mainly for the Department of Agriculture and the Customs Service. He was an old-fashioned patriot who questioned the scientific establishment and government constantly.

Mark was intrigued by nature's anomalies for most of his life. For almost sixty years, he actively pursued historical records and eyewitness testimonies of cryptozoological phenomena. He traveled extensively throughout the Americas—and the world.

I first corresponded with Mark when author, television nature show host, zoologist, Fortean, and cryptozoologist Ivan T. Sanderson introduced us through letters late in the 1960s. Mark served as the director of the Sanderson-founded Society for the Investigation of the Unexplained (SITU) in the early 1970s, right before Sanderson's death in 1973.

When we first started conversing, Mark was living in Minnesota and I was living in Illinois. Before long, we would visit each other and engage in long conversations about unknown hominoids, cryptozoology, and our latest theories.

On his first visit to my home in Decatur, Illinois, Mark and I talked into the wee hours of the morning. After that, we exchanged thoughts on a regular basis via visits, letters, and phone conversations.

Mark actively researched mystery cats, hairy hominoids, surviving anthropoids, ancient civilizations, and hundreds of other topics. He had a large library of books and files. Before he died, he donated large boxes full of original research materials to the ICM. He wanted his research to find a home where future generations could learn more from his work.

Mark wrote Fortean articles and a column for *Fortean Times*. He edited and published the journal *Wonders*, which was chiefly devoted to cryptozoology and other related topics that interested him.

Some of Mark's works, such as *Thunderbirds; Natural Mysteries; The Yeti, Bigfoot & True Giants*; and *Living Fossils*, were self-published. He saw a couple of his books more formally published in recent years, such as *Thunderbirds: America's Living Legends of Giant Birds*.

Mark was a fearless theorist. He proposed that North America is home to the Bigfoot of the Patterson-Gimlin footage, as well as different primates, such as the True Giant (which he believed was *Gigantopithecus*) and the Taller-hominid (which he saw as survivors of the recorded fossil known as *Homo gardarensis*). I assisted him in finishing the book *True Giants: Is* Gigantopithecus *Still Alive?* in 2010.

Mark and I kept track of the Minnesota Iceman throughout the Midwest from 1968 to 1969. I interviewed Mark about his involvement with the Minnesota Iceman for my "Afterword" in the book *Neanderthal: The Strange Saga of the Minnesota Iceman*. Mark also appeared on the show *Unsolved Mysteries* detailing his Minnesota Iceman involvement. The case was featured as a part of the September 25, 1994, episode on NBC and then repeated on the Lifetime Channel for decades.

Mark had an incredible memory for details and his insights always impressed me. In 1999, I coined the name "Marked Hominids" to honor him. He had invented the "Taller-hominid" concept, which was poorly understood in the field. I rechristened his "Taller-hominids" as "Marked Hominids" to describe this specific type of Bigfoot-like creatures seen from Siberia to the Eastern United States. Often noted as aggressive and pie-bald ("marked") in coloration, these are not the Neo-Giants, the Bigfoot/Sasquatch so well-known from the Pacific Northwest. Because Mark had first described them as separate from Bigfoot, it seemed natural to christen them "Marked Hominids."

Mark and I appeared together on *Coast to Coast A.M.* with George Noory to discuss Mothman and Thunderbirds in 2005. I presented an overview of the Mothman case. We discussed how the creature is similar to a giant owl and may use air turbulence from cars to facilitate its flight. Mark joined the show to share material on Thunderbirds, which reportedly having a wingspan of eighteen to twenty feet—twice the size of any known birds. He recounted the 1977 Lawndale, Illinois, case, where the giant bird picked up and briefly carried a young boy. Though unharmed, the boy's hair turned gray after the incident. Mark said there were similar instances in the 1800s.

Mark occasionally did interviews on his own, like a *Coast to Coast A.M.* show with Ian Punnett in 2006. He talked about Giant Owls, Thunderbirds, and other mysterious behemothic birds. He said Native American legends and contemporary accounts talk about such birds, particularly in the West Virginian Appalachian Mountains. Some of these "great owls" are reportedly man-sized, with wingspans of ten feet. Mark contended they were probably the inspiration behind the Mothman stories of the 1960s. He coined the term "Bighoot" to describe the true origins of the Mothman sightings, which he related to giant owls.

George Noory invited him back on July 20, 2006, to discuss Thunderbirds and other mysterious creatures. During this discussion, Mark related the Lizardman to folklore tales of mermen and mermaids. Lizardmen, Mark posited, are seven-foot tall, scaly creatures that inhabit the water, yet somehow can remove their fish parts when they arrive on land.

He also said they sometimes mate with humans. On the topic of unknown hominids in the Yukon Territory and Alaska, he shared modern-day reports of Neanderthals seen wearing crude clothing and carrying axes.

British vertebrate paleontologist Darren Naish referred to both Mark's "Bighoot" theory (and some of mine as well) as "speculative creature building." Mark was not easily dissuaded and thought the data would eventually speak for itself.

Mark proposed a new technique for studying cryptids, which he termed "telebiology." He formalized his thoughts about the concept in print almost twenty-five years ago. When discussing this technique in relation to the search for unknown primates, he wrote:

> With temporary captives, we should do the best we can with them and then set them free. The results will be genuine knowledge in the records we will then have, and we will have invested in the future of a new relationship with our primate relatives. This approach is part of what I have called 'telebiology,' a means by which we can begin to study the cryptids that have been the object of cryptozoology. If we make the effort to study animals at a distance, using our brains and technology, we can succeed where others have failed in the past. If we can accept that starting to study a species with a dead animal can be difficult, then we can put that goal at the end of the process instead of making it a requirement to do anything at all.[1]

This is the essence of Mark's earlier, now unavailable, self-published *Lizardmen* book that serves as the foundation for this revised and expanded edition, published for the ICM. Mark became one of the first honorary members of the International Cryptozoology Society early in 2016.

I will miss Mark deeply. This world is much less interesting without him. But thankfully, he has left words, thoughts, and books for all of us to ponder into the future.

Loren Coleman
Bangor, Maine
March 26, 2023

---

1 Hall, *The Yeti, Bigfoot & True Giants*, 110.

Appendix 2
The Mermaid Motif at the Start of the 20th Century

Charles M. Skinner

IN DIME MUSEUMS AND COUNTY FAIRS, one may still find among the "attractions" a mermaid, dried and stuffed, consisting of the upper half of a monkey artlessly joined to the lower half or two-thirds of codfish, the monkey's head usually adorned with a handful of oakum or horse-hair. When this kind of thing was first exhibited by the lamented P. T. Barnum, it is just possible that some bumpkin really believed it to be a mermaid, but the invention has become so common of late that it is found in the curio-shops of every town, and as an eye-catching device is often put into show-cases by some merchant who deals in anything but mermaids. Trite and ridiculous as this patchwork appears, it symbolizes a belief of full three thousand years. Men have always been prone to fill with imaginations what they have never sounded with their senses, and it is to this tendency we owe poetry and the arts. The sea was a mystery, and is so still. It was easy to people its twilight depths with forms of grace and beauty and power, for surely the denizens taken from it were strange enough to warrant strange beliefs.

And so the old faith in men and women who lived beneath the water was passed down from generation to generation, and from race to race, changing but little from age to age. Ulysses stopped the ears of his crew with wax that they should not hear the sirens luring them toward the rocks as his ship sailed by, and knowing the magic of their song had himself bound to the mast, so, hearing the ravishing music, he might not escape if he would. In a later day we hear of the Lorelei singing on her rock, striking chords on her golden harp, and, as the raptured fisherman steered close, with eyes filled by her beauty and ears by her music, he had a moment's consciousness of a skull leering at him and harsh laugher clattering in echoes along the shore; then his boat struck and filled, and the dark flood curtained off the sky. Wagner has made familiar the legend of the Rhine daughters, singing impossibly under the river as they swim about the reef of gold—treasure stolen by the gnome, Alberich, who in that act brought envy, strife, greed, and injustice into the world, and accomplished the destruction of the gods themselves. The wild tales of Britain and Brittany, of thefts and revenges by the sea-creatures, are among the oldest of their myths, and when we cross to our side of the sea, the ocean people are close in our wake and they follow us through the fresh waters and far out in the Pacific.

Among the Antilles, as in the South Seas, the tritons blow their conchs and shake their shaggy heads, while the daughters of the deep gather, at certain seasons, on the water, or about some favorite rock, and sing. Always, in Eastern versions of the myth, there is music, save in the case of Melusina, who became a half fish only on Saturdays, when her husband was supposed not to be watching, and this music follows the myth around the world.

Triton from *Testacea Musei Caesarei Vindobonensis* by Ignaz von Born, 1780.

Among the vague traditions of certain Alaskan Indians is one of an immigration from Asia, under lead of "a creature resembling a man, with long, green hair and beard, whose lower part was a fish; or, rather, each leg a fish." He charmed them so with his singing that they followed him, unconsciously, and reached America.

We find in Canada the tale of a dusky Undine, a soulless water sprite, who, through love of a mortal, became human. Some of the beings of the sea were of more than human power and authority, gods in fact; barbarian Neptunes. Such was the Pacific god, Rau Raku, who, being entangled in a fishing-net, was lugged to the surface, sputtering tremendously. Yet he had no grudge against the fisherman. That trembling unfortunate was too small for his revenge. He would devastate the whole earth to which he had been thus unceremoniously dragged, and, bidding his captor take himself away while he made trouble, he deluged the globe until all upon it had perished, except the fish, the fisherman, and a few land animals that the sole human survivor had taken to a lofty island with him.

The mermaid of story was a damsel fair to view, until she had risen from the waves so as to show her fish-like ending. It was her habit to sit on sunny beaches, comb her golden hair with a golden comb, and sing delightfully, though her wilder sisters would perch on juts of rock on lonely islands and scream in frightening ways when a gale was coming. When the sea maidens went ashore they sometimes met sailors and fishermen, and if they liked these strangers a frank avowal of love was made; for it is always leap year in the ocean. It was a most uncomfortable position for a mortal to be placed in, especially one who had a wife waiting for him at home, because if their addresses were rejected the mermaids were liable to throw stones, and always with fatal results; or they would brew mists, and set loose awful storms; yet, if the man who inspired this affection was not coy, and yielded to one of these slippery denizens, she dragged him under the sea forthwith, unless he could persuade her to compromise on a cave or a lonely rock as a home, for it is reputed that mortals have formally wedded them and raised amphibious families.

On the Isle of Man they tell of one caught in a net, who was woman to the waist and fish as to the rest of her. As she sulked in captivity, refusing to eat or speak—perhaps they forgot to offer raw fish for her supper—it was decided to let her escape; and as she wriggled over the beach she was heard to tell her people, as they arose to greet her, that the earthmen did nothing wonderful except to throw away water in which they had boiled eggs!

The home of the mermaids was at the bottom of the deep. A diver, who said he had reached it, reported a region of clear water, lighted from below by great, white stones and pyramids of crystal. These haunts contained bowers of coral, gardens of bright sea weeds and mosses, tables and chairs of amber, floors of iridescent shell and pearls, gems strewn about the jasper grottoes—diamonds, rubies, topazes—and the sea people had combs and ornaments of gold. Columbus was disappointed in the mermaids that he saw in the Caribbean. They were not, to his eyes, so handsome as the romancers had alleged, nor were their voices sweet. The doubters claim that he was asleep when the mermaids appeared, and that he saw nothing but the sea cow, or manatee, which is neither tuneful or pretty.

Reprinted from Charles M. Skinner, *Myths & Legends of Our New Possessions & Protectorate* (Philadelphia: Lippincott, 1900), 78–83.

Appendix 3
Along the Trek: A Popular Summary
Loren Coleman

CHAPTER TEN OF JOHN A. KEEL'S *Strange Creatures from Time and Space* is devoted entirely to a hodgepodge of primarily American beasts that Keel felt were related to the Abominable Snowman and Bigfoot.[1] These ape-like monsters, said to be roaming the "hollers and hills" of the eastern and southern United States, have always been a focus of my fieldwork research efforts. One-third of Keel's cases for his chapter actually came from my files. When I picked up the book in the 1970s and read it for the first time, something about Keel's grouping made me uncomfortable. Finally, it dawned on me that Keel's chapter title was not quite right. His title, "Creatures from the Black Lagoon," was a very catchy Keelian way of noting something strange was slithering through the swamps. Still, the original movie monster he had used as the model was anything but ape-like.

The eponymous Creature from the Black Lagoon was a reptilian "Lizardman," a bipedal, human-sized beast that looked more like a spiny-skinned were-lizard than a hairy were-gorilla or werewolf. The thing was literally a half-man, half-amphibious reptile.

Now, as creatures go, these types of monsters are rare but by no means nonexistent. So much "monster lumping" occurs among Forteans and cryptozoologists that my focus became examining the accounts of existing creatures closely, whether called Bigfoot, swamp monsters, or whatever, to reveal the true Creatures from the Black Lagoon, the Lizardmen.

West Coast Cases

The confusion between various uncommon cases comes into focus when you examine the history of one classic so-called "Bigfoot" story, the Wetzel sighting in Riverside, California, in the fall of 1958. The Riverside sighting entered public awareness so soon after the Bluff Creek,

The California original "Bigfoot" print find by Jerry Crew in October 1958 has often been retold in Sasquatch books, beginning with Ivan T. Sanderson's 1961 *Abominable Snowmen.*[2] In 1982, I tracked down and interviewed Charles Wetzel and his family personally and found some interesting new information.

Wetzel was driving his green, two-door 1952 Buick Super near Riverside, California, when he saw "it." Saturday, November 8, 1958, is a night Charlie would not soon forget. He even remembers which radio station (KFI in Los Angeles) he had tuned into that night. Wetzel neared the section of North Main Street where the Santa Ana River infrequently overflows its banks. Sure enough, water was rushing across the pavement at a spot where the road dips. So Charles slowed down. Within moments he was struck by two sensory events, which caught him off guard. First, his car radio started to transmit lots of static. He changed stations, he told me, but to no avail. Next, he saw what he thought was a temporary danger sign near the flooded site. Before he could think twice about any of this, Charles Wetzel saw a six-foot creature bound across his field of vision and stop in front of his Buick. The wire services widely circulated the account starting on November 10, 1958.

The creature had a "round, scarecrowish head like something out of Halloween," Wetzel said at the time. He described it then, and again to me later, as having no ears or nose, a beak-like, protuberant mouth, and fluorescent, shining eyes. The skin was "scaly, like leaves, but definitely not feathers," Wetzel recalled during our 1982 talk.

The creature was waving "sort of funny" with incredibly long arms. It appeared to be walking from the hips, almost like it had no knees. Wetzel then remembered another detail not noted at the time: the legs stuck out from the sides of the torso, not from the bottom. The gurgling sounds it made were mixed with high-pitched screams. It reached across the hood when it saw Wetzel and began clawing at the windshield. Terrified, Wetzel grabbed the .22 High Standard pistol he kept in the car because he was often on the road at night. Clutching the gun, the frightened Californian stomped on the gas. "Screeching like a f--r," as Wetzel graphically put it, the creature tumbled forward off the hood and was run over by the car. Wetzel could hear it scrape the pan under the engine, and later, police lab tests revealed that something had indeed scrubbed the grease from the Buick's underside.

The police used bloodhounds to search the area, but the dogs found nothing, and the officers were left with only the sweeping claw marks on Wetzel's windshield to ponder. Then, the very next night, a black something jumped out of the underbrush near the same site and frightened another motorist. In recent years, sightings of the strange three-toed Bigfoot have been reported from surrounding areas of southern California. Most notably, the smelly eight-footer, dubbed "Buena Foot," was seen emerging from a Brea Creek drainage ditch in Buena Park in May 1982.[3]

Clearly, the Wetzel creature fits the reptilian mold better than the anthropoid. A thing with fluorescent eyes, a protuberant mouth, and a body covered with scales certainly doesn't sound like a Sasquatch. And as Mark noted previously in this book, fluorescence occasionally appears among our mysterious aquatic species.

The connection to water is a strong theme in all of these Lizardmen accounts. Unsurprisingly, the next piece of the puzzle comes from the lake monster file. Trekking up the west coast, we find the following report concerns a monster that actually looks like it stepped out of the wardrobe room of the Black Lagoon movie.

Thetis Lake is a small manmade reservoir near Colwood, British Columbia. It is not far from Victoria. Cadboro Bay, off Victoria and Vancouver Island, and well known for the perennial sea monster "Cadborosaurus." So understandably, a new creature in the neighborhood would be grouped under the same façade by the press. But the Thetis Lake Monster appears to be something else altogether.

On August 19, 1972, Gordon Pike and Robin Flewellyn said a five-foot-tall animal appeared on the surface of Thetis Lake and chased them from the beach. Flewellyn was cut on the hand by six razor-sharp points atop the monster's head. A Royal Canadian Mounted Police (RCMP) officer was quoted at the time as saying, "The boys seem sincere, and until we determine otherwise, we have no alternative but to continue our investigation."[4]

The following Wednesday afternoon, August 23, the Thetis Monster was encountered again. Mike Gold and Russell Van Nice said they saw "it" around 3:30 PM on the other side of the lake, away from the recreation area of its first appearance. Mike Gold noted: "It came out of the water and looked around. Then it went back in the water. Then we ran!"

He described the creature as "shaped like an ordinary body, like a human being body but it had a monster face, and it was all scaly [with] a point sticking out its head [and] great big ears. It was silver."

The RCMP investigated the Thetis Monster "because it's been reported to us, and we have to check these things out." There was no resolution. A similar creature has been reported in nearby Puget Sound in Washington. The Kwakiutl Indian merman, *Pugwis*, is sort of a cross between Sasquatch and the Creature from the Black Lagoon. Its fishlike face and paired incisors make this undersea spirit a prominent figure in Indian legend and is easily recognized in wood-carved art.

On To the Great Lakes

The Thetis Monster also resembles one seen in Saginaw, Michigan, in 1937. A man-like monster climbed up a riverbank, leaned against a tree, and then returned to the river. The fisherman who witnessed this appearance suffered a nervous breakdown. Perhaps, this Saginaw tale, combined with the reports of clawed and three-toed prints from Wisconsin to Missouri and other supposedly "Bigfoot" or "manimal" encounters, should be re-examined in light of the reptilian Creatures from the Black Lagoon.

For example, the big Deltox Swamp, Wisconsin flap investigated by Ivan T. Sanderson in 1969 has always been shelved with Bigfoot reports. But what are we to make of the creature's tracks, which were said to be like footprints of "a good-sized man with swim fins"?[5]

This trait is more frequently reported in the so-called eastern Bigfoot reports than most people realize. For example, there's the case of the Charles Mill Lake creature of Mansfield, Ohio, where I visited and re-interviewed locals in 1975. A green-eyed, seven-foot, seemingly armless humanoid was seen late in March 1959 by Michael Lane, Wayne Armstrong, and Dennis Patterson.[6] It emerged from the lake and left behind "tracks that resembled the footgear worn by skin divers." The thing was seen again in 1963 and described as luminescent, with eyes "shining like a cat's."[7] My examination of the site of these encounters concludes that Charles Mill Lake's swampy affinities would undoubtedly be a good home for a Black Lagoon beast.

In the annals of midwestern monster hunters, few discuss a cryptic chapter as they talk about the fieldwork done in Louisiana, Missouri. In the early 1970s, many folks looked into the widespread reports of "Momo" (Missouri Monster). One investigator was traveling down a back road and was surprised to see what appeared to be a grown man dressed in complete skin-diving gear down to the swim fins. This was miles from any logical skin-diving site. Does this report of a "frogman" have something to do with our inquiry?

Or what are we to make of reports of some "things" merging from Mothman-like beasts into the Creatures from the Black Lagoon?

In Wisconsin, for example, there is the 1990s story of the "Reptile Man" of Highway 13. Fortean researcher Richard Hendricks chronicled the case on his lamentably defunct website but fortunately summarized the encounter in his *Weird Wisconsin* book.[8] The highway goes north to south, down the middle of the state. A Department of Natural Resources warden was traveling down 13, south of Medford, Wisconsin, when suddenly he saw a figure standing in the middle of the road. It was hominoid and about the same size as a human but appeared to be green-scaled. Incredibly, as the warden grew closer to the thing, he saw wings pop from the beast's back. Then, suddenly, it went straight up and flew over his vehicle, coming down and landing behind him. Later, the warden discovered that he wasn't the only person to have seen this thing. He learned that some highway workers had an almost identical experience along this same section of the road.

Though this quest is pulling us to the U.S. Midwest, the focus of the accounts, let's deal first with matters in New Jersey and New York.

Mid-Atlantic Swamps

When charted, the reports, tales, and sightings of Creatures from the Black Lagoon form a continuous watery line down the Susquehanna River. It continues through the so-called Southern Tier of New York State, along the Delaware River, and ends in the counties of Morris and Sussex in New Jersey. During the summer of 1973, residents of the Newton-Lafayette area in New Jersey described a giant, man-like alligator they had seen locally. Newspaper reporters wrote about an old Indian tale from the region that told of a giant, man-sized fish that could never be caught. In 1977, Alfred Hulstrunk, a naturalist with New York's State Nature and Historical Preserve Trust, reported, "from the Tamarack swamps along the Delaware, a scaled, man-like creature appears at dusk from the red, algae-ridden waters to forage among the fern and moss-covered uplands." [9]

The Ohio River Valley

The New York-New Jersey record for such creatures is minimal compared to the overwhelming series of narratives from the Ohio River Valley.

During the 1960s and 1970s, I dug into the back issues of the Louisville, Kentucky, *Courier-Journal*. I discovered a gem that has kept me pondering its meaning for decades. The interesting little item appeared in the October 24, 1878 issue. A "Wild Man of the Woods" was allegedly captured in Tennessee and then placed on exhibit in Louisville. The creature was described as six feet, five inches tall, with eyes twice the average size. His body was "covered with fish scales."[10] This 1878 article would make some sense in the context of these Lizardmen, assuming it wasn't some unfortunate individual with a skin abnormality.

Almost a hundred years later, again near Louisville, there are more stories of reptilian entities. In October 1975, near Milton, Kentucky, Clarence Cable reported a "giant lizard" roaming the forests near his automobile junkyard. Anomalist author Peter Guttilla described the creature Cable surprised as "about fifteen feet long, [with] a foot-long forked tongue, and big eyes that bulged something like a frog's. It was dull-white with black-and-white stripes across its body with quarter-size speckles over it."[11]

Reports by other Trimble County, Kentucky witnesses, which were investigated on-site by Mark A. Hall, confirmed the "giant lizard" sightings.[12] Trimble County's northern boundary is the Ohio River, and upriver from here, even stranger stories abound.

On August 21, 1955, near Evansville, Indiana, Mrs. Darwin Johnson was almost pulled forever into the depths of the Ohio River. In what seems to have been a very close encounter with one of these creatures, Mrs. Johnson of Dogtown, Indiana, was swimming with her friend Mrs. Chris Lamble. They were about fifteen feet from shore when suddenly, something grabbed her from under the surface. It felt like the "hand "had huge claws and "furry" (or scaly?) palms. It appeared behind her, grabbed her left leg, gripped her knee, and pulled her under. She kicked and fought herself free. It pulled her under again. Although both women could not see the thing, they screamed and yelled to scare it away. Finally, Mrs. Johnson lunged for Mrs. Lamble's inner tube, and the loud "thump" apparently scared "it" away as "it" released its grip. Back on shore, Mrs. Johnson received treatment for the scratches and marks on her leg.[13] Fortean investigator Terry Colvin passed on the information that Mrs. Johnson had a palm-print-shaped green stain below her knee that could not be removed for several days.

Interestingly, Colvin learned that an individual who identified himself as an Air Force colonel visited the Johnsons. He took voluminous notes and warned them not to talk further about the incident. Of course, this sounds very similar to a "Man-In-Black" encounter.

For anyone who has seen the film *The Creature from the Black Lagoon*, the Ohio River encounter of Mrs. Johnson is already familiar. Her attack was foreshadowed in that movie. John Baxter, in his *Science Fiction in the Cinema*, recounts those moments:

> "A key scene of the film is when the heroine (Julie Adams) enters the water for a swim, unaware that the creature is swimming just below her, admiring. Shots looking towards the surface show the girl penetrating a Cocteau-like mirror, her white suit with its accentuated breasts, her choreographed leg movements all overtly sexual. Gliding beneath her, twisting lasciviously in a stylized representation of sexual intercourse, the creature, his movements brutally masculine and powerful, contemplates his ritual bride, though his passion does not reach its peak until the girl performs some underwater ballet movements, explicitly erotic poses that excite the Gill-Man to reach out and clutch at her murmuring legs."[14]

According to Baxter, the cinematic Lizardman, the first Creature of the Black Lagoon, is presented in the film "as a force of elemental power, not maliciously evil but 'other-directed,' a fragment of a world where our ideas of morality have no relevance."

We can only speculate that the same may be true of the very "real" Creatures from the Black Lagoon.

Loveland's Trolls and Frogman
From Evansville, Indiana's watery attack to Loveland, Ohio's case of the "trolls under the bridge," the story continues along the Ohio River Valley, just down the river from Point Pleasant and Mothman country.

Leonard H. Stringfield's inquiry into the "affair under the bridge" is perhaps well known to most ufology readers but not many others.[15] The sighting occurred in March 1955 at 4:00 AM in Branch Hill, Ohio. Robert Hunnicutt, a businessman, saw three man-like "trolls" kneeling on the side of the road. They were about three feet tall, had gray skin, and seemed to be wearing tight-fitting gray clothes or had skin that appeared so. They had frog-like faces, long slender arms, and normal eyes but no eyebrows. One held a dark object (emitting blue flashes) between its raised arms. Hunnicutt tried to approach them but "must have lost consciousness" because he found himself driving to the police station without remembering what happened. An FBI investigation followed, and a guard was placed on the bridge.

Stringfield discovered that, clearly, some weird things were happening in the region at the time. On August 25, 1955, in Winton Woods, Ohio, four teens in an automobile saw a luminous creature standing near a fire hydrant. And six weeks earlier, on July 3, 1955, in Stockton, Georgia, Mrs. Wesley Symmonds was driving near this town when she saw four "bug-eyed" creatures near the road. They were small beings with thin arms, large eyes, and pointed chins. Two were turned away from the witness, one was bending over with something like a stick in its hand, and the fourth was facing her with its right arm raised. It had bulging eyes, some sort of cap, no visible mouth, a long pointed nose, a chin that came to a sharp point, and long thin arms with claws.[16]

The famed Kelly Creatures case, often mislabeled as the Hopkinsville, Kentucky Incident, took place during this same time frame, on August 21-22, 1955. The Sutton family and their relatives had to deal with a literal invasion of three-foot-tall "goblins" that glowed silver, had big eyes, and oversized heads. The things were hit by gunfire by the rural folks who were scared but firm in their resistance. The "goblins" merely floated away when shot.[17]

The 1955 incidents in Indiana, Georgia, and Kentucky seem less isolated in the context of what was happening in Ohio in 1955. But there is one point I would make about the Hunnicutt encounter in Branch Hill, Ohio, that for years may have misled those interested in such matters: the famous drawing of the "trolls under the bridge," with their lopsided chests, was sketched by Stringfield based on his impression of what was seen and was not drawn under the direction of the witness.[18] Instead, these frog-like "trolls" more appropriately belong in Fortean creature chronicles than in UFO books, as does an incident that took place 17 years later in nearby Loveland.

In March of 1972, on two separate occasions, two Ohio policemen saw what has become known as the "Loveland Frogman." The incidents were investigated by Ron Schaffner and Richard Mackey, who interviewed the officers.

The first incident occurred on a clear, cold night at 1:00 AM, March 3, 1972. Officer Ray Shockey was en route to Loveland via Riverside Road when he thought he saw a dog in a field on Twightwee Road beside the road. But then the "dog" stood up, its eyes illuminated by the car lights, looked at him for an instant, turned, and leapt over a guardrail. Shockey saw it go down an embankment into the Little Miami River, a mere fifteen or so miles from the Ohio River. He described the thing as weighing about sixty pounds, measuring about three to four feet tall, and having textured leathery skin and a face like a frog or a lizard. Shockey drove to the police station and returned with Officer Mark Matthews to look for evidence of the creature. They turned up scrape marks leading down the small hill near the river.

On St. Patrick's Day, March 17, 1972, Officer Matthews was driving outside Loveland when he had a similar experience. Seeing an animal lying in the middle of the road, he stopped to remove what he thought was a dead critter. Instead, when the officer opened his squeaky car door, the animal got up into a crouched position like a football player. The creature hobbled to the guardrail and lifted its leg over the fence, keeping an eye on Matthews the whole time.

Perhaps it was the funny smirk on its face, but Matthews decided to shoot it. He missed, however, probably because the thing didn't slow down. Matthews later said his impression was that the creature stood more upright than how Shockey had described it. One area farmer told investigators he saw a large frog or lizard-like creature during the same month as the officers' sightings.

The reports of the 1972 Frogman had been hard on the witnesses. "Those two officers took a lot of flack about the sightings back then," said a local businessman who wished not to be identified in a 1985 newspaper story about the sightings. "People made fun of them and the city."

Years later, in 1999, during interviews, Mark Matthews explained that he was tired of talking about the "Frogman" and that he had seen an iguana missing its tail. But at the time, both witnesses saw something like an upright man-like lizard about 4 feet tall. And then there is the matter of the sketch. Officer Shockey's sister drew it for them shortly after their experience with the creature. It clearly is a giant frogman, a bipedal creature. Then, in 2001, I asked investigator Ron Schaffner about Matthews' recent attempts to pull back from his original story. Schaffner told me: "Why, after all these years, is Matthew debunking the story? I'm not sure. Could be a number of reasons. But both officers told us that it resembled the sketch in 1976. Why would they show us a composite drawing of this creature back in 1976 and tell us that it looked like the drawing? I lived in Loveland for about five years and the story still circulates with many variations. Just maybe Matthews is tired of hearing the story and all the variations."[19]

Those who witnessed similar creatures in 1955 were apparently unavailable as a safety net for those police officers who saw something in 1972. The ridicule curtain came down hard on them. Can we blame anyone for wishing it never happened to them?

The Lizardman Mystery

"The trouble in trying to understand all reported monsters," said Charles Fort, "is their mysterious appearances and disappearances."

How true.

Can we really conceptualize that Lizard Beings are real? No wonder people toy with the idea of explaining one mystery with another, that we are seeing visitors that exist on another plane, for example. John Keel believes that Mothman, other glowing-eyed beasties, and Creatures from the Black Lagoon are nothing more than demonological elementals, here to cause trouble. In a world gone mad, perhaps he's on to something. With such a wild set of reports, we must think outside the box. So what is going on here?

During the 1980s Dale Russell, a Canadian paleontologist, promoted the idea that reptiles had as much chance to evolve into an intelligent, bipedal form as mammals.[20] His drawings of this upright, reptilian, intellectual animal look suspiciously like the composite picture from the sightings of the Ohio Valley/Black Lagoon creatures. Are these beasts future time travelers lost in some time/space warp? Or are they infrequent visitors from an ultraterrestrial plane? Or have these stepped out of flying saucers? Or do you feel more comfortable with the idea of a breeding population of scaly, man-like, upright creatures lingering along the edges of some of America's swamps-cryptids of the upright reptilian variety? Merbeings among us?

Are Lizardmen just contemporary versions of the mermen, mermaids, and merbeings, the water creatures of the past? Mark A. Hall certainly felt this may be the case. Perhaps, the mermaids and mermen of ancient lore are still being seen today, although he thinks in far lesser numbers. Yet this group of water-connected beings ranges far beyond the merpeople of yore and includes such varieties as the scaly-looking but perhaps misnamed Lizardmen or the fiery-eyed Latino phenomenon known as the Chupacabras. So maybe the answer dwells more in the water than on the land.

Something is out there. That's for sure.

An early version of this essay appeared in Loren Coleman's *Mothman and Other Curious Encounters* (2002).

ENDNOTES

1 Keel, *Strange Creatures from Time and Space* (1976).
2 Sanderson, Abominable Snowmen (1961).
3 "Police have no clue to smelly 'big foot'" *Santa Ana Orange County Register*, May 12, 1982.
4 "Thetis Monster Seen by Boys," *Times Colonist* (Victoria, BC), August 22, 1972.
5 Sanderson, "Wisconson's 'Abominable Snowman.'" *Argosy* (April 1968).
6 "Boys Report Seeing Green-Eyed Monster." *Mansfield News Journal*, 28 March 1959.
7 Gaynor, Donn. "'Hairy Monster' Has Folks in a Tizzy," *Mansfield News Journal*, 27 July 1963.
8 Godfrey and Hendricks, *Weird Wisconsin*, 2005.
9 Hulstrunk, "Assorted Ghosts, Ghouls and Goblins of New York State," *Binghamton Press and Sun-Bulletin*, August 14, 1977.
10 "Wild Man of the Woods," *Courier-Journal* (Louisville, KY), October 24 1878.
11 Guttilla, *The Bigfoot Files*, 2003, 167.
12 Hall, *Natural Mysteries*, 1991.
13 "Woman Battles 'It' in Ohio River." *Evansville Press* (Evansville, IN). August 15, 1955.
14 Baxter, *Science Fiction in the Cinema*, 1970, 121.
15 Stringfield, *Situation Red: the UFO siege!*, 1977, 87-92.
16 Stringfield, 87.
17 "Story of Space-Ship, 12 Little Men Probed Today," *Kentucky New Era*, 22 August 1955.
18 Stringfield, 91.
19 Legate, "Officer who shot 'Loveland Frogman' in 1972 says story is a hoax." *WCPO*, August 05, 2016.
20 Russell and Séguin "Reconstruction of the small Cretaceous theropod Stenonychosaurus inequalis and a hypothetical dinosauroid." *Syllogeus* 37 (1982).

Bibliography

Preface—From Legend to Reality

"$25,000 Offered for Mermaid." *The Daily Colonist (Victoria BC)*, 15 June 1967.

Gee, Henry. "Flores, God and Cryptozoology." *Nature* (27 October 2004). https://doi.org/10.1038/news041025-2

Hall, Mark A. "Why Nothing Gets Solved in One Lifetime," *Wonders* 8 (2), June 2003.

"Mermaid Legend Taking Shape." *The Daily Colonist (Victoria BC)*, 14 June 1967.

"Mermaid Visits New Spot." *The Daily Colonist (Victoria BC)*, 23 June 1967.

"Short Sight Sizzler." *The Daily Colonist (Victoria BC)*, 13 June 1967.

Introduction: Wading Into the Swamp with Mark

Hall, Mark A. *Living Fossils: The Survival of Homo Gardarensis, Neandertal Man, and Homo Erectus.* Wilmington, NC: Mark A. Hall Publications, 1999.

Hall, Mark A. *Lizardmen: The True Story of Mermen and Mermaids.* Wilmington, NC: Mark A. Hall Publications, 2005.

Hall, Mark A. *Thunderbirds: America's Living Legends of Giant Birds.* New York: Paraview Press, 2004.

Hall, Mark A. *The Yeti, Bigfoot & True Giants: An Introduction.* Minneapolis, MN: Mark A. Hall Publications, 2005.

Hall, Mark A., and Loren Coleman. *True Giants: Is* Gigantopithicus *Still Alive?* San Antonio, TX: Anomalist Books, 2010.

Williams, Marcel Francis. "Cranio-dental evidence of a hominin-like hyper-masticatory apparatus in Oreopithecus bamboo," *Bioscience Hypotheses* 1 (3) 2008.

Chapter 1—The Lure of Mermaids

Bassett, Fletcher S. *Legends and Superstitions of the Sea and of Sailors in All Lands and at All Times.* Chicago: Belford, Clarke & Co., 1885.

Benwell, Gwen, and Arthur Waugh. *Sea Enchantress: The Tale of the Mermaid and Her Kin.* NY: Citadel, 1965.

Broadhurst, C. Leigh, et al. "Littoral Man and Waterside Woman: The Crucial Role of Marine and Lacustrine Foods and Environmental Resources in the Origin, Migration, and Dominance of Homo sapiens." *EurekaSelect*, 2011.

Delson, "An Anthropoid Enigma: Historical introduction to the study of Oreopithecus bambolii." *Journal of Human Evolution* 15 (November 1987).

Gosse, Philip Henry. *Romance of Natural History*. 2nd series. London: James Nisbet and Co., 1861.

Hall, Mark A. *Living Fossils: The Survival of Homo Gardarensis, Neandertal Man, and Homo Erectus*. Wilmington, NC: Mark A. Hall Publications, 1999.

"Satyrs and Centaurs," *Wonders* 10 (3), September 2006.

"The Satyrs in Our Midst," *Wonders* 9 (3), September 2005.

"Satyrs in Africa," *Wonders* 10(3), September 2006.

Hardy, Alister. "Was Man More Aquatic in the Past?" *New Scientist* 7 (174), 17 March 1960.

Hürzeler Johannes. *"Zur systematischen Stellung von Oreopithecus." Verhandlungen der Naturforschenden Gesellschaft in Basel* 65 (1) 1954.

Lawson, John Cuthbert. *Modern Greek Folklore and Ancient Greek Religion: A Study in Survivals*. Cambridge: University Press, 1910.

Sanderson, Ivan T. *"Things"* and *More "Things."* Kempton, IL: Adventures Unlimited Press, 2006.

Scribner, Vaughn. *Merpeople: A Human History*. London: Reaktion Books, 2020.

Verhaegen, M. J. B. "The Aquatic Ape Theory: Evidence and a possible scenario," *Medical Hypotheses* 16 (1), January 1985.

Verhaegen, Marc, and Stephen Munro, "Pachyosteosclerosis suggests archaic Homo frequently collected sessile littoral foods." *Homo - Journal of Comparative Human Biology* 62 (4), August 2011.

Vaneechoutte, Mario (ed) et al. *Was Man More Aquatic in the Past?* Sharjah, United Arab Emirates: Bentham Science Publishers, 2018.

Chapter 2––Oreopithecus

Coleman, Loren, and Patrick Huyghe. *Field Guide to Bigfoot and Other Mystery Primates*. San Antonio, TX: Anomalist Books, 2015.

Culotta, Elizabeth. "Spanish Fossil Sheds New Light on the Oldest Great Apes," *Science* 306 (5700), 19 November 2004.

Fleagle, John G. *Primate Adaptation and Evolution.* 2 ed. San Diego: Academic Press, 1999.

Goksör E., et al. "Bradycardic response during submersion in infant swimming." *Acta Paediatrica* 91 (3), March 2002.

Hapgood, Charles H. *Earth's Shifting Crust.* NY: Pantheon Books, 1958. Reprinted in a revised edition as *The Path of the Pole.* Philadelphia: Chilton Book Company, 1970.

Kennedy, G.E. "The relationship between auditory exostoses and cold water: a latitudinal analysis." *American Journal of Biological Anthropology* 71 (4) December 1986.

Kirschvink, Joseph L., et al. "Evidence for a Large-Scale Reorganization of Early Cambrian Continental Masses by Inertial Interchange True Polar Wander." *Science* 277 (5325), 25 July 1997.

Kreichgauer, Damian. *Die Äquatorfrage in der Geologie.* Steyl, Netherlands: Missionsdruckerei, 1902.

Moyà-Solà, Salvador, et al. "*Pierolapithecus catalaunicus,* a New Middle Miocene Great Ape from Spain." *Science* 306 (5700), 19 November 2004.

Quick, Nicola J. et al. "Extreme diving in mammals: first estimates of behavioural aerobic dive limits in Cuvier's beaked whales," *Journal of Experimental Biology* 223 (18), September 2020.

Rutledge, Archibald. "Mysteries of Nature: peculiar occurrences that defy explanation." *Nature Magazine* 27 (5) May 1936.

Wolfe, Hannah Wondy. "The Great Shaking" *Flying Saucer Review* 40 (2), Summer 1995.

Young, Emma. "The Beast With No Name." *New Scientist* 184 (2468), 09 October 2004.

Chapter 3—Native Accounts of North America

Benwell, Gwen, and Arthur Waugh. *Sea Enchantress: The Tale of the Mermaid and Her Kin.* NY: Citadel, 1965.

Boas, Franz. "Central Eskimo." in *Sixth Annual Report of the Bureau of Ethnology.* Washington DC: Government Printing Office, 1888.

Bushnell, Jr., David I. *The Choctaw of Bayou Lacombe, St. Tammany Parish, Louisiana.* Bureau of American Ethnology Bulletin 48. Washington: GPO, 1909.

Gatschet, Albert. "Water Monsters of the American Aborigines," *Journal of American Folklore* 12 (47), October-December 1899.

Haupt, Herman. [Papers concerning North American Indians]. Edward
    E. Ayer Manuscript Collection; Newberry Library, Chicago, IL.
    manuscript. Ayer MS 366.
Hawk, Kevin. "Apotamkin" in *American Myths, Legends, and Tall
    Tales,* volume 1. ed. Christopher R. Fee and Jeffrey B. Webb. Santa
    Barbara, CA: ABC-CLIO, 2016.
Krukoff, Martin, "Aboriginal Ghost, yítcanam mayayú, Creek Man." in
    "Stories, Myths and Superstitions of Fox Island Aleut Childen." ed.
    Jay Ellis Ransom. *Journal of American Folklore* 60 (235), January-
    March 1947.
 Merriam, C. Hart. *The Dawn of the World - Myths and Weird Tales Told
    by the Mewan Indians of California.* Cleveland, OH: Arthur H.
    Clark Company, 1910. Reprinted as *The Dawn of the World: Myths
    and Tales of the Miwok Indians of California.*
Orr, R. H. *Sailing directions for Nova Scotia, Bay of Fundy, and South
    Shore of Gulf of St. Lawrence.* Washington, DC: Government
    Printing Office, 1891.
Packard, Christopher. *Mythical Creatures of Maine.* Camden, Maine:
    Down East Books, 2021.
Reagan, Albert B. "Picture Island," *The Southern Workman* 55 (.10),
    October 1926. Reprinted in *The Minnesota Archaeologist* 22(2), Fall
    1958.
Rose, Carol. *Giants, Monsters, and Dragons: An Encyclopedia of Folklore,
    Legend, and Myth.* Santa Barbara, CA: ABC-CLIO, 2000.
Skinner, Charles M. *American Myths and Legends.* Vol. 2. Philadelphia:
    J.B. Lippincott Co, 1903.
"Superstitions of the Passamaquoddies." *Journal of American Folklore*
    2(6), July-September 1889.
Wallis, Wilson D., and Ruth Sawtell Wallis. *The Micmac Indians of
    Eastern Canada.* Minneapolis: University of Minnesota Press,
    1955.
Wilson, Herbert Earl. *The Lore and the Lure of the Yosemite; The Indians,
    Their Customs, Legends and Beliefs, Big Trees, Geology, and the Story
    of Yosemite.* San Francisco, CA: Sunset Press Pub, 1922.
Winchell, N. H. *The Aborigines of Minnesota.* St. Paul, MN: The Pioneer
    Company, 1911.

Chapter 4—North America

Hall, Mark A. "October 1958 in the History of Bigfoot." *Wonders* 9(3), September 2005.

Josselyn, John. *An Account of Two Voyages to New-England*. London: Giles Widdows, 1674.

"Little Monster Breaks Fishing Poles of Women." *Northwest Arkansas Times (Fayetteville)*, 28 May 1973.

"Lizard in Lake?" *The Province* (Vancouver, BC), 26 August 1972.

"A Mermaid in Lake Superior." *The Canadian Magazine and Literary Repository* 11 (2), May 1824.

"A Mermaid in the Susquehanna." *The York (PA) Daily*, 08 June 1881.

"Merman Seen by Fish Boat Crew off Port." *San Pedro News-Pilot* (CA), 25 May 1935.

"Nondescript are These Creatures, Which Resemble a Human Being in Many Ways." *Cincinnati Enquirer*, 06 September 1894.

"A Sea Nymph Seen at Garbarus." *Saint John Colonist* (Newfoundland), 09 July 1886.

"A Strange Animal in Florida." *Brooklyn Daily Eagle* (NY), 01 October 1882.

Sutton, David. "Fishing trip results in sighting of 'Big Foot'?" *News Courier* (Athens, AL), 08 August 1978.

"Thetis Monster Seen by Boys." *Times Colonist* (Victoria, BC), 22 August 1972.

Whitbourne, Richard. *A Discovrse and Discovery of Nevv-Fovnd-Land*. London: Felix Kyngston, 1620.

Chapter 5—South America and the Caribbean

Bacchilega, Cristina, and Marie Alohalani Brown, eds. *The Penguin Book of Mermaids*. NY: Penguin Books, 2019.

Barco Centenera, Martín del. *Argentina y Conquista del Rio de la Plata*. Buenos Aires: A. Estrada, 1912. [rep of 1602 ed].

Bastide, Roger. *The African Religions of Brazil*. trans. Helen Sebba. Baltimore: The Johns Hopkins University Press, 1978.

"Bog Walk gorge is Haunted – Residents say mermaids, spirits live there." *The Weekend Star* (Kingston, Jamaica), 15 December 2017.

Braham, Persephone. "Song of the Sirenas: Mermaids in Latin America and the Caribbean" in *Scaled for Success: The Internationalisation of the Mermaid*. ed. Philip Hayward. New Barnet, UK: John Libbey Publishing Ltd, 2018.

Cappick, Marie. "Key West Resident Sees a Mermaid." *Paths* 1 (5), July 1934.

Cardim, Fernao. *Tratados da Terra e Gente do Brasil*. Rio de Janeiro: J. Leite & Cia, 1925. [rep of 1583 ed].

Cavada, Francisco Javier. *Chiloé y los Chilotes*. Santiago, Chile: Imprenta Universitaria, 1914.

Chisholm, C. *An Essay on the Malignant Pestilential Fever, Introduced into the West Indian Islands from Boullam, on the Coast of Guinea, as it appeared in 1793, 1794, 1795, and 1796,* vol. 2, rev. ed. London: J. Mawman, 1801.

Clarke, Jay. "New roads show Haiti's side not seen in cities." *Chicago Tribune*, 13 February 1977.

Colnett, James. *A Voyage to the South Atlantic and round Cape Horn into the Pacific Ocean*. London: W. Bennett, 1798.

Conzemius, Eduard. *Ethnographical Survey of the Miskito and Sumu Indians of Honduras and Nicaragua*. Bureau of American Ethnology Bulletin 106. Washington: GPO, 1932.

Couto de Magalhães, José Vieira. *O Selvagem,* part II. Rio de Janeiro: Typographia da Reforma, 1876.

Gândavo, Pêro de Magalhães de. *História da Província Santa Cruz*. ed. Clara C. Souza Santos, and Ricardo M. Valle. São Paulo: Hedra, 2008. [rep of 1576 ed].

Goudsward, David. *Sun, Sand, and Sea Serpents*. San Antonia, TX: Anomalist Books, 2020.

Grubb, W. Barbrooke. *A Church in the Wilds*. ed. H. T. Morrey Jones. NY: E. P. Dutton and Co., 1914.

"He Did See a Mermaid." *Brooklyn Daily Eagle*, 07 October 1894.

Holmquist, Ulla, and Carole Fraresso. *Machu Picchu and the Golden Empires of Peru*. Viterbo, Italy: Laboratoriorosso, 2021.

Léry, Jean de. *Histoire d'un Voyage fait en la Terre du Brésil: dite Amérique*. A La Rochelle: Antoine Chuppin, 1578.

Mathison, Ruddy. "Flat Bridge haunted – Residents claim mermaid lives in Rio Cobre." *The Star (Kingston, Jamaica)*. 09 June 2021, http://jamaica-star.com/article/news/20210609/flat-bridge-haunted-residents-claim-mermaid-lives-rio-cobre

"Woman killed in Bog Walk Gorge crash." *The Gleaner* (Kingston, Jamaica). 9 June 2021.

Métraux, Alfred. Myths of the Toba and Pilagá Indians of the Gran Chaco, *Memoirs of the American Folklore Society*, Vol. XL. Philadelphia: American Folklore Society (1946).

Meurger, Michel, and Claude Gagnon. *Lake Monster Traditions: a crosscultural analysis.* London: Fortean Tomes, 1988.

"Mythical maidens come to life in the Bahamas," Winnipeg *Free Press*, 1 March 1975.

"Philadelphia, April 29." *The Pennsylvania Gazette*, April 22-29, 1736.

Rosales, Diego de. *Historia General de el Reyno de Chile*, vol 1. Valparaiso, Chile: Imprenta de Mercurio, 1877. [rep of 1674 ed].

Schmidt, Bettina E. "Mermaids in Brazil: The (ongoing) creolisation of the water goddesses Oxum and Iemanjá" in *Anthropology and Cryptozoology: Exploring Encounters with Mysterious Creatures.* ed. Samantha Hurn. London: Routledge, 2016.

Chapter 6—Europe

Almqvist, Bo. "Of Mermaids and Marriages. Seamus Heaney's 'Maighdean Mara' and Nuala Ní Dhomhnaill's 'an Mhaighdean Mhara' in the Light of Folk Tradition." *Béaloideas - The Journal of the Folklore of Ireland Society* 58 (1990).

Carrington, Richard. *Mermaids and Mastodons: A Book of Natural and Unnatural History.* NY: Rinehart & Company, 1957.

Coggeshall, Radulphi de. *Chronicon Anglicanum.* ed. Josephus Stevenson. London: Longman and Co., 1875.

Holmberg, Uno. "Finno-Ugric, Siberian" in *Mythology of All Races* (1927), vol.4 (rep 1927 edition. NY: Cooper Square Publishers, Inc. 1964.

Hudson, Henry. "A second Voyage or Employment of Master Henry Hudson, for finding a passage to the East Indies by the North-east." *Collections of the New-York Historical Society* v.1 (1809) NY: I. Riley (1811).

Kidal, Simon, "Efterretning om Finners og Lappers hedenske religion," *Det skandinaviske Litteraturselskabs Skriftert* III (2), 1807.

Leather, Ella Mary. *The Folk-Lore of Herefordshire.* London: Sidgwick and Jackson, 1912.

"A Live Mermaid. And No Mistake," *Londonderry Standard (Northern Ireland)*, 5 September 1838.

"Mermaid" *Galway Advertiser (Ireland)*, 25 September 1819.

Morvan, F. *Legends of the Sea*. NY: Crescent Books, 1980.

Pliny the Elder, *Pliny's Natural History. In Thirty-seven Books*. Vol.1 (rep. 1601 edition), trans. Philémon Holland. [London]: Wernerian Club, 1847.

Scribner, Vaughn. *Merpeople – A Human History*. London: Reaktion Books, 2020.

Simpson, Jacqueline, and Steve Roud. *A Dictionary of English Folklore*. Oxford: Oxford University Press, 2000.

Teit, J. A. "Water-Beings in Shetlandic Folk-Lore, as Remembered by Shetlanders in British Columbia." *The Journal of American Folklore* 31(120) April-June 1918.

[Vigo merman]. *Universal Spectator and Weekly Journal* (London, England), 5 May 1739.

Weddell, James. *A Voyage towards the South Pole, performed in the years 1822-24*. London: Longman, Rees, Orme, Brown, and Green (1827).

Chapter 7—Asia

"圓珠院所蔵「人魚のミイ」研究最終報告 (Final report on research on the "mermaid mummy" in the Enju-in collection)." Okayama, Okayama Prefecture, Japan: Kurashiki University of Science and the Arts, 7 February 2023.

Batchelor, John. *The Ainu and Their Folk-Lore*. London: The Religious Tract Society, 1901.

Bernard, Penny S. "Mermaids, Snakes and the Spirits of the Water in Southern Africa: Implications for River Health." Presented at a Conference on River Management in Southern Africa. Rhodes University, Grahamstown, 2000.

Caidin, Martin. *Natural or Supernatural? A Casebook of True Unexplained Mysteries*. Chicago: Contemporary Books, 1993.

"China 'Mermaid descendants' weave garments from fish skin." *The Asahi Shimbun* (Osaka, Japan) [English ed.], 21 January 2020.

Dennys, N.B. *The Folk-Lore of China*. Hong Kong: China Mail Office, 1876.

Du Halde, Jean-Baptiste. *Description Géographique, Historique, Chronologique, Politique et Physique de l'empire de la Chine et de la Tartarie Chinoise*. Vol. IV. La Haye: Chez Henri Scheurleer (1763).

Fan Duanang. *Yue Zhong Jian Wen*. Sanshui]: Wu Dian Zhai, [1801]. rep. Guangzhou: Guangdong Gaodengjiaoyu Chubanshe, 1988.

Furukawa, Yuki, and Rei Kansaku. "Amabié—A Japanese Symbol of the COVID-19 Pandemic." *Journal of the American Medical Association* 324 (6), 11 August 2020.

Garifdjanov, Rafic. "Mysterious amphibious creature of the Caspian Sea." *Pravda* (Moscow, Russia) [English ed.], 24 March 2005.

Hayward, Philip. "The Mermaidisation of the Ningyo" in *Scaled for Success: The Internationalisation of the Mermaid* (2018).

Holmberg, Uno. "Finno-Ugric, Siberian" in *Mythology of All Races* (1927), vol.4 (rep 1927 edition). NY: Cooper Square Publishers, Inc. 1964.

"Killed a Mermaid," *Galveston Daily News*, 28 Sept. 1891. *Tokio Shimpo* article, newswire.

Kiyoaki Sato "Genkou Zenkoku Youkai Jiten" *Hogen Sosho* 7, Okayama: Nihon Minzoku Gakkai (1935).

Kunio Ozawa. "Scientists try to unravel mystery of eerie 'mermaid mummy.'" *The Asahi Shimbun* (Osaka, Japan) [English ed.], 19 February 2022.

"Tall tale: Study finds 'mermaid mummy' largely a molded object," *The Asahi Shimbun* [English ed.], 8 February 2023.

Marks, William. *I Saw Ogopogo!* Peachland, BC: W. Marks, 1971.

Mayers, W. F. "Mermaids and Mermen in the Chinese Seas." *Notes and Queries on China and Japan* 3, no. 7, July 1869.

"Mermaid Found at Last." *New York Tribune*, 10 July 1911.

"Mermaid Found in San Francisco Bay," *Sausalito News* (CA), 29 June 1907.

"'Monster' Seen in Lake," *Daily Colonist* (Victoria, BC), 22 August 1972.

Nagano Eishun. "Amabiko the prophetic beast, revisited" in *The Forefront of the Study of Yōkai Culture*. ed. Komatsu Kazuhiko. Tokyo: Serika Shobō, 2009.

"Nihongi: Chronicles of Japan from the Earliest Times to A.D. 697." trans. W. G. Aston. *Transactions and Proceedings of the Japan Society* 1, Supplement 1, 1896.

Pietsch, Theodore W. "Samuel Fallours and his 'Sirenne' from the Province of Ambon" *Archives of Natural History* 18 (1), February 1991.

Sacchini, Francesco. *Historiæ Societatis Iesu pars secunda siue Lainius.* Antwerp: Martini Nutii, 1620.

Sadamatsu, Shinjiro. "'Mermaid bones' from 13th century keep legend alive in Fukuoka." *The Asahi Shimbun* (Osaka, Japan) [English ed.], 13 February 2017.

Sanderson, Ivan T. "Luminous People and Others." *Pursuit* 6 (3), July 1973.

Ssu-ma Ch'ien. *The Grand Scribe's Records.* trans. William H. Nienhauser, Jr. volume 1. Bloomington: Indiana University Press, 1994.

Swancer, Brent. "Orang Ikan," *CryptoZooNews*, 17 July 2009. http://www.cryptozoonews.com/orang-ikan/

Valentijn, François. *Beschryving van Oud en Nieuw Oost-Indië.* vol. 3. Amsterdam: Gerard onder de Linden, Dordsecht, J. Van Braam, 1724.

Vaneechoutte, Mario,et al (ed). *Was Man More Aquatic in the Past?* ed. Sharjah, United Arab Emirates: Bentham Science Publishers, 2018.

Varner, Gary R. *Creatures in the Mist.* NY: Algora Publishing, 2007.

Zhang Chenliang. *Chinese National Geography - Notes of Hai Cuo Tu.* Beijing: CITIC Press (2016).

Chapter 8—Africa

Al-Ahmadi, Fahd Amer. "Houryaat al-Bahr Bayin al-Khoraafa wa al-Asl." *Al-Riyadh*, 12 May 2005.

Alexander, James Edward. *Narrative of a Voyage of Observation among the Colonies of Western Africa.* London: Henry Colburn, 1874.

Aschwanden, Herbert. *Karanga mythology. An analysis of the consciousness of the Karanga in Zimbabwe.* trans, Ursula Cooper. Gwero, Zimbabwe: Mambo Press, 1989.

Bernard, Penny S. "Ecological Implications of Water Spirit Beliefs in Southern Africa: The Need to Protect Knowledge, Nature, and Resource Rights" in *Science and Stewardship to Protect and Sustain Wilderness Values: Seventh World Wilderness Congress Symposium, 2001.* ed. Alan Watson and Janet Sproull. Ogden, UT: U.S. Department of Agriculture, Forest Service, Rocky Mountain Research Station, 2003.

"Mermaids, Snakes and the Spirits of the Water in Southern Africa: Implications for River Health." Presented at a Conference on River Management in Southern Africa. Rhodes University, Grahamstown, 2000.

The Book of The Thousand Nights and a Night. trans. R. F. Burton. ed
  Leonard C. Smithers. London: H. S. Nichols, 1897.
"Chipata Overspill Residents Burn Down Church After Suspecting a
  Mermaid Inside." *The Zambian Observer*, 13 January 2020.
Cotter, Holland. "From the Deep, a Diva with Many Faces." *The New
  York Times*, 2 April 2009.
Dapper, O. *Naukeurige Beschrijvinge der Afrikaensche Gewesten.*
  Amsterdam: Jacob van Meurs, 1668.
de Maillet, Benoît. *Telliamed: or, Discourses between an Indian
  Philosopher and a French Missionary, on the Diminution of the
  Sea, the Formation of the Earth, the Origin of Men and Animals.*
  London: T. Osborne, 1750.
Drewal, John Henry. "Interpretation, Invention, and Re-presentation in
  the Worship of Mami Wata." *Journal of Folklore Research* 25 (1-2),
  January-August 1988.
Füreri ab Haimendorf, Christophori. *Itinerarium Ægypti, Arabiæ,
  Palæstinæ, Syriæ, aliarumque regionum orientalium.* Norimberge
  [Germany]: Abrahami Wagenmanni, 1620.
How, M. H. *The Mountain Bushmen of Basutoland.* Pretoria, South
  Africa: Van Schaik, 1970.
Ittmann, Johannes. "Der Kultische Geheimbund Djĕngú an Der
  Kameruner Küste." *Anthropos* 52 (1-2), 1957.
Leeuwenburg, J. "A Bushman Legend from the George District. *South
  African Archaeological Bulletin* 25 (99-100), December 1970.
"'Mermaid' Spotted on Kiryat Yam, Beach." Israel Hayom 12 August
  2009. Quoted and translated into English by *Israel National News.*
  https://www.israelnationalnews.com/flashes/169395
"Mermaids stopping Govt work: Sipepa Nkomo." *The Herald* (Harare,
  Zimbabwe), 30 January 2012.
"Mystery surrounds boat tragedy." *The Herald* (Harare, Zimbabwe), 28
  September 2010.
Orpen, J. M. "A Glimpse into the Mythology of the Maluti Bushmen."
  The Cape Monthly Magazine 9 (July 1874). Reprinted as "Folklore
  of the Bushmen." *Folklore* 30 (2), 30 June 1919.
Osoba, Funmi. *Benin Folklore A Collection of Classic Folktales and
  Legends.* London: Hadada Books, 1993.
Page, A. B. "San Mermaids in Lesotho." *South African Archaeological
  Bulletin* 33 (127), June 1978.

Pekeur, Aldo. "Mysterious 'Mermaid' Rises from the River." *Independent Online News*, 16 January 2008. https://www.iol.co.za/news/south-africa/mysterious-mermaid-rises-from-the-river-385945

Pontoppidan, Erik. *The Natural History of Norway.* trans. Andreas Berthelson. London: A. Linde, 1755.

Puckler Muskau, Prince. *Egypt Under Mehemet Ali.* London: Henry Colburn Publisher, 1845. Originally published in German as *Semilasso in Afrika : aus den Papieren des Verstorbenen.* Stuttgart: Hallberger, 1836.

Shalaby, Manal. "The Middle Eastern Mermaid" in *Scaled for Success: The Internationalisation of the Mermaid.* ed. Philip Hayward. New Barnet, UK: John Libbey Publishing Ltd, 2018.

Shuker, Karl. *The Beasts That Hide from Man: Seeking the World's Last Undiscovered Animals.* NY: Paraview Press, 2003.

Siegel, Brian. "Water Spirits and Mermaids: The Copperbelt *Chitapo.*" in *Sacred Waters: Art for Mami Wata.* ed. Henry J. Drewal. Indianapolis: Indiana University Press, 2008.

Thevet, *Les Singularitez de la France Antarctique.* Paris: Maurice de La Porte, 1557.

Thevet, Andre. *The Nevv Found World; or, Antarctike.* trans Thomas Hacket. (London: Henrie Bynneman, 1568).

Varner, Gary R. *Creatures in the Mist.* NY: Algora Publishing, 2007.

Verbeek, Léon. *Filiation et Usurpation: histoire socio-politique de la région entre Luapula et Copperbelt.* Tervuren, Belgium: Musée royal de l'Afrique centrale, 1987.

"What! Saw a Real Mermaid?" *Norfolk* (NE) *News-Journal*, 13 January 1911. Distributed nationally.

"Zimbabwe mermaids appeased at pumphouse." *news24* (South African Press Association), 12 February 2012. https://www.news24.com/News24/Zimbabwe-mermaids-appeased-at-pumphouse-20120212

Chapter 9—Oceania

Alexander, Skye. *Mermaids: The Myths, Legends, & Lore.* Avon, MA: Adams Media, 2012.

Archey, Gilbert. "Evolution of Certain Maori Carving Patterns." *Journal of the Polynesian Society* 42 (3) #167, September 1933.

Bacchilega, Cristina, and Marie Alohalani Brown, eds. *The Penguin Book of Mermaids.* NY: Penguin Books, 2019.

Brosses, Charles de. "VIII – Diego Hurtado et Fernand de Grijalva en Polynésie." *Histoire des Navigations aux Terres Australes.* (Paris: Chez Durand), 1756.

"Article VIII – Diego Hurtado to Polynesia." *Terra Australis Cognita; or, Voyages to the Terra Australis, or Southern Hemisphere, during the Sixteenth, Seventeenth, and Eighteenth Centuries,* vol I. trans John Callander. (Edinburgh: A. Donaldson) 1766.

Clark, Jerome. *Encyclopedia of Strange and Unexplained Physical Phenomena.* Detroit, MI: Gale Research, 1993.

Clark, Jerome. *Unexplained!* Chicago, IL: Visible Ink Press, 1998.

Cutchin, Joshua. *The Brimstone Deceit.* San Antonio, TX: Anomalist Books, 2016.

Dominy, Nathaniel J., and Peter W. Lucas. "Ecological importance of trichromatic vision to primates." *Nature* 410, no. 6826, 2001.

Gilad, Yoav, et al. "Loss of olfactory receptor genes coincides with the acquisition of full trichromatic vision in primates." *PLoS biology* 2, no. 1, 2004. https://doi:10.1371/journal.pbio.0020005

Gislén, Anna, and Erika Schagatay. "Superior Underwater Vision Shows Unexpected Adaptability of the Human Eye." in *Was Man More Aquatic in the Past?* Sharjah, United Arab Emirates: Bentham Science Publishers, 2018.

Green, Felicity. *Togart Contemporary Art Award* (exhibition catalogue). Darwin, Northern Territory: Toga Group, 2007.

"Half-human-half-fish Beauty," *Eastern Horizon* 17 (9), September 1978.

Herrera y Tordesillas, Antonio de. *Historia General de los Hechos de los Castellanos en las Islas y Tierra Firme del Mar Océano* (Madrid: Juan de la Cuesta 1615). decada V, libro VII, capitulo III, IIII.

McLintock, A. H. *An Encyclopaedia of New Zealand.* Wellington, New Zealand: R.E. Owen, 1966.

Meurger, Michel, and Claude Gagnon. *Lake Monster Traditions: a crosscultural analysis.* London: Fortean Tomes, 1988.

Poignant, Roslyn. *Oceanic mythology: the myths of Polynesia, Micronesia, Melanesia, Australia.* London: Paul Hamlyn, 1967.

Smith, Gordon. "Cryptozoology: Seeking Mermaids, Living Legends, Mythical Monsters." *Chicago Tribune,* 27 January 1985.

Wagner, Roy. "Further Investigations into The Biological and Cultural Affinities of The Ri." *Cryptozoology – Interdisciplinary Journal of the International Society of Cryptozoology* 2 (1983).

"The *Ri* – Unidentified Aquatic Animals of New Ireland, Papua New Guinea." *Cryptozoology – Interdisciplinary Journal of the International Society of Cryptozoology* 1 (1982).

Wavehill, Ronnie. "Karukany (Mermaids)." in *Yijarni: True Stories from Gurindji Country.* ed. Erika Beatriz Charola and Felicity Helen Meakins. Canberra, Australia: Aboriginal Studies Press, 2016. Wavehill's article is reprinted in Bacchilega and Brown, *The Penguin Book of Mermaids,* 2019.

Whitfield, Anna. "Mermaid tales appear in myths around the world — Arnhem Land included." *Late Night Live* (Australian Broadcasting Corporation Radio National), 10 July 2018.

Williams, Thomas R. "Identification of the Ri through Further Fieldwork in New Ireland, Papua New Guinea." *Cryptozoology – Interdisciplinary Journal of the International Society of Cryptozoology* 4, 1985.

Winkler, Lawrence. *Stories of the Southern Sea.* Victoria, BC: First Choice Books, 2016.

The Wugularr Aboriginal Community with Liz Thompson. *The Mermaid and Serpent.* Port Melbourne: Harcourt Education, 2010.

Yamazaki, Kunio, et al. "Odortypes: Their Origin and Composition." *Proceedings of the National Academy of Sciences,* 96 (4):1522-5, March 1999. DOI:10.1073/pnas.96.4.1522.

Chapter 10—The Mysterious Ones

Baldwin, Anne Elizabeth. "Menehune," in *Storytelling: An Encyclopedia of Mythology and Folklore.* ed. Josepha Sherman. Armonk, NY: Sharpe Reference, 2011.

Balzano, Christopher. *Dark Woods: Cults, Crimes, and the Paranormal in the Freetown State Forest, Massachusetts.* Atglen, PA: Schiffer Publishing, 2008.

Bates, Russell. "Legends of the Kiowa" *INFO Journal,* no. 52, May 1987.

Bernal, Ignacio. *The Olmec World.* Berkeley: University of California Press 1969.

Bierhorst, John. *The Deetkatoo: Native American Stories about Little People.* NY: William Morrow and Company, 1998.

"Brewster." *Provincetown Advocate,* July 16, 1873.

Carmichael, Alexander. *Carmina Gadelica.* Edinburgh: T. and A. Constable, 1900.

Coleman, Loren. *Monsters of Massachusetts: Mysterious Creatures in the Bay State.* Mechanicsburg, PA: Stackpole Books, 2013.

Dorsey, George A., and Alfred L. Kroeber. *Traditions of Arapaho.* Field Colombian Museum, Publication 81, Anthropological Series V. Chicago: Field Museum of Natural History, 1903.

Foster, Michael Dylan. *The Book of Yōkai.* Oakland: University of California Press, 2015.

Fox, Luke. *North-west Fox, or, Fox from the North-West passage.* London: B. Alsop and Tho. Fawcet, 1635.

Gardiner, John. Alligators in the Bahamas, *Science* ns 8(194), 22 October 1886.

Goldstein, Lorrie. "Tunnel monster of Cabbagetown?" *Sunday Star* (Toronto, ON), 25 March 1979.

Hall, Manly P. "The Mystery of the Thunderbird," *Overland Monthly and Out West Magazine* 87 (4) April 1929.

Hamilton, Robert. *Mammalia: Amphibious Carnivora.* The Naturalist's Library, vol. XXV: Animals vol. XI. ed. William Jardine. London: Henry G. Bohn, [c.1840].

Heuvelmans, Bernard. *On the Track of Unknown Animals.* London: Rupert Hart-Davis, 1959.

Mallery, Garrick. *Picture-writing of the American Indians.* NY: Dover Publications, 1972. The work originally appeared in 1894 as pp. 3-807 of the *Tenth Annual Report of the Bureau of Ethnology to the Secretary of the Smithsonian Institution, 1888-89.*

Marcot. Bruce G. *Owls of Old Forests of the World.* Portland, OR: U.S. Dept. of Agriculture, Forest Service, Pacific Northwest Research Station, 1995.

"Mermaid." *The American Journal of Science and Arts* II (2), November 1820.

"A Mermaid at Lewes," *Daily Gazette (Wilmington, DE),* 5 January 1880.

"The Mermaid's Grave, Nunton, Benbecula (S Uist parish): putative grave-marker." In *Discovery & Excavation in Scotland - An Annual Survey of Scottish Archaeological Discoveries, Excavation and Fieldwork.* ed. Colleen E. Batey. Edinburgh, Scotland: Council for Scottish Archaeology, 1994. https://doi.org/10.5284/1000284

Moseley, Mary. *The Bahamas Handbook.* Nassau: Nassau Guardian, 1926.

Sanderson, Ivan T. "Traditions of Submen in Arctic and Subarctic North America." *Pursuit* 16 (1), 1983.

Scherman, Katharine. *Spring on an Arctic Island*. Boston, MA: Little, Brown and Company, 1956.

Speck, Frank G. *Naskapi, the Savage Hunters of the Labrador Peninsula*. Norman: University of Oklahoma Press, 1935.

"Story of Space-Ship, 12 Little Men Probed Today," *Kentucky New Era (Hopkinsville, KY)*, 22 August 1955.

"This monster may be a hoax, then again..." *The Boston Globe*, 16 May 1977.

Wetmore, Alexander. "Bird remains from cave deposits on Great Exuma Island in the Bahamas." *Bulletin of the Museum of Comparative Zoology* 80(12), October 1937.

Witthoft, John, and Wendell S. Hadlock. "Cherokee-Iroquois Little People." *The Journal of American Folklore* 59 (234), October-December 1946.

Chapter 11—1988—The Year of the Lizardman

Coleman, Loren, and Patrick Huyghe. *Field Guide to Bigfoot and Other Mystery Primates*. San Antonio, TX: Anomalist Books, 2015.

D'Anghiera, Pietro Martire. *De Orbe Novo, The Eight Decades of Peter Martyr d'Anghera*. Vol. 2. trans. Francis Augustus MacNutt. NY: G. P. Putnam's Sons, 1912.

Drape, Joe. "Lizard Man a Tall Tail? Spooky Swamp Legend Slithers Into Limelight." *Atlanta Constitution* (Atlanta, Georgia), 31 July 1988.

"Fame Follows Close Encounter of the Lizard Kind." *Charlotte (NC) Observer*, 2 August 1988.

Georgas, George "'Lizard' calls besiege Lee sheriff." *The Item (Sumter, SC)*, 21 July 1988.

Hall, Mark A. "Guide to North American Monsters," *Wonders* 6 (1), March 1999.

"Pinky, the Forgotten Dinosaur." *Wonders* 1 (4), December 2002.

Horswell, Cindy. "On a scale of one to 10, it rates a downright scary 11." *Houston Chronicle*, 31 July 1988.

"The Lizard Man returns; seen by 'respected' citizen." *The Item* (Sumter, SC), 2 September 1988.

"'Lizard Man' sightings rouse some excitement." *Anderson Independent-Mail* (Anderson, SC), 21 July 1988.

Peterson, Patrick. "Lizardman has Coast connection." *Sun Herald* (Biloxi, MS), 13 August 1988.

Squires, Chase. "Lizard-on-the-lam feared in area." *The Index-Journal* (Greenwood, SC), 25 July 1988.

Stein, Theo. "Bigfoot believers - Legitimate scientific study of legend gains backing of top primate experts." *Denver* (CO) *Post,* 5 January 2003.

"Teen Who Spotted Lizard Man Reaping Benefits." *The Aiken Standard* (Aiken, SC), 2 August 1988.

Tuten, Jan. "First Lizard Man spotter passes polygraph." *The State* (Columbia, SC), 3 September 1988.

"Monster bash – 'Reputable people' say they saw the elusive 'Lizard Man.'" *The State* (Columbia, SC), 20 July 1988.

"Pet alligator not monster, sheriff says." *The State* (Columbia, SC), 21 July 1988.

"'Skunk ape' still causing a stink." *The State* (Columbia, SC), 27 July 1988.

"Tracks pique public interest." *The State* (Columbia, SC), 15 August 1988.

"Trooper convinced that 'something is out there.'" *The State* (Columbia, SC), 28 July 1988.

Chapter 12—Lizardmen of the Carolinas

"Bayard's Leap" *Lincolnshire Notes & Queries* 7, 1902-1903.

Blevins, J. N. "Stories the Goose Hunters Never Hear," *The State* 35 (10), 15 October 1967.

Bolden, Bria. "Savannah woman wakes up to car severely damaged by unknown animal." *WTOC* (Savannah, GA). https://www.wtoc.com/2021/09/11/savannah-woman-wakes-up-car-severely-damaged-by-unknown-animal/

Brown, Frank C. "The Headless Woman" *Frank C. Brown Collection of North Carolina Folklore,* vol. 1. Durham, NC: Duke University Press, 1952.

Cantrell, Bertie. "The Legend of Slippery Hill," *The State* 49 (1), April 1982.

Eiichirô, Ishida. "The Kappa Legend," *Folklore Studies* 9 (1950).

Fisher, Douglas. "The Four-Mile Desert: A Horror Story," *North Carolina Folklore Journal* 23 (1), February 1975.

Fowler, Malcolm. "A Tour of Cape Fear," in *The Story of Fayetteville and the Upper Cape Fear* by John A. Oates. Charlotte, NC: The Dowd Press, 1950.

Glendinning, William. *The Life of William Glendinning, Preacher of the Gospel.* Philadelphia: W.W. Woodward, 1795.

"Hunt Animal Half Snake Half Woman." *Philadelphia Inquirer (PA),* 21 August 1907.

McNeill, Robert E. "Legends from the Devil's Stairs." *North Carolina Folklore Journal* 26 (3), November 1978.

"Queer Things in Georgia." *Atlanta Constitution* (GA), 20 April 1895.

Shropshire Folk-Lore; A Sheaf of Gleanings. ed. Charlotte Sophia Burne London: Trübner & Co., 1883.

Smith, Elizabeth Simpson. "The Ghost that Rides," *North Carolina Folklore Journal* 23 (2), May 1975.

Tuten, Jan. "Lizardless Summer – Scaly swamp man, tourists elude Lee County." *The State* (Columbia, SC), 24 July 1989.

"An Uncanny Monster." *The Larned Eagle-Optic* (Larned, Kansas), 24 June 1892. Multiple reprints.

Afterword—What Can We Learn?

Burns, J. W. "Introducing B. C.'s Hairy Giants: A collection of strange tales about British Columbia's wild men as told by those who say they have seen them." *Macleans,* 1 April 1929.

Eberhart, George M. *Monsters: A Guide to Information on Unaccounted for Creatures, Including Bigfoot, Many Water Monsters, and Other Irregular Animals.* New York: Garland, 1983.

Roth, John E. *American Elves: An Encyclopedia of Little People from the Lore of 380 Ethnic Groups of the Western Hemisphere.* Jefferson, NC: McFarland, 1997.

Strain, Kathy Moskowitz. *Giants, Cannibals & Monsters - Bigfoot in Native Culture.* Surrey, BC: Hancock House Publishers, 2008.

Appendix 1—In Memory of Mark A. Hall by Loren Coleman

Hall, Mark A. *Lizardmen.* Mark A. Hall Productions, 2005.

Hall, Mark A. *Living Fossils.* Mark A. Hall Productions, 1999.

Hall, Mark A. *Natural Mysteries.* Mark A. Hall Productions, 1989.

Hall, Mark A. *The Yeti, Bigfoot & True Giants.* Mark A. Hall Productions, 1994.

Hall, Mark A. *Thunderbirds,* Mark A. Hall Productions, 1988.

Hall, Mark A. *Thunderbirds: America's Living Legends of Giant Birds.* Paraview Press, 2004.

Heuvelmans, Bernard. *Neanderthal: The Strange Saga of the Minnesota Iceman*. Anomalist Books, 2016.

Appendix 2—The Mermaid Motif at the Start of the 20th Century by Charles M. Skinner

Skinner, Charles M. *Myths and Legends of our New Possessions and Protectorate*. Philadelphia: J.B. Lippincott Company, 1900.

Appendix 3—Along the Trek: A Popular Summary by Loren Coleman

Arnold, Jack, director. *Creature from the Black Lagoon*. Universal Pictures, 1954. 1 hr., 19 min.

Baxter, John. *Science Fiction in the Cinema*. NY: A.S. Barnes & Co, 1970.

"Boys Report Seeing Green-Eyed Monster." *Mansfield News Journal*, 28 March 1959.

Coleman, Loren. *Mothman and Other Curious Encounters*. New York: Paraview, 2002.

Gaynor, Donn. "'Hairy Monster' Has Folks in a Tizzy." *Mansfield News Journal*, 27 July 1963.

Godfrey, Linda S., and Richard D. Hendricks. *Weird Wisconsin*. NY: Barnes & Noble Publishing, 2005.

Guttilla, Peter. *The Bigfoot Files*. Santa Barbara, CA: Timeless Voyager Press, 2003.

Hall, Mark A. *Natural Mysteries: Monster Lizards, English Dragons, and Other Puzzling Animals*. Minneapolis, MN: Mark A. Hall Publications and Research, 1991.

Hulstrunk, Alfred. "Assorted Ghosts, Ghouls and Goblins of New York State." *Susquehanna* (Sunday supplement, *Binghamton Press and Sun-Bulletin*), 14 August 1977.

Keel, John A. *Strange Creatures from Time and Space*. London: Sphere, 1976.

Leggate, James. "Officer who shot 'Loveland Frogman' in 1972 says story is a hoax." *WCPO* (Cincinnati, OH). https://www.wcpo.com/news/local-news/hamilton-county/loveland-community/officer-who-shot-loveland-frogman-in-1972-says-story-is-a-hoax

"Police have no clue to smelly 'big foot'" *Santa Ana Orange County Register*, 12 May 1982.

Russell, D. A., and R. Séguin "Reconstruction of the small Cretaceous theropod Stenonychosaurus inequalis and a hypothetical dinosauroid." *Syllogeus* 37 (1982).

Sanderson, Ivan T. *Abominable Snowmen: Legend Comes to Life.* NY: Cosimo Classics, 2008. rep. of 1961 ed.

"Wisconson's 'Abominable Snowman.'" *Argosy* 31(4), April 1968.

"Story of Space-Ship, 12 Little Men Probed Today," *Kentucky New Era* (Hopkinsville, KY), 22 August 1955.

Stringfield, Leonard H. *Situation Red: The UFO Siege!* Garden City, NY: Doubleday, 1977.

"Thetis Monster Seen by Boys." *Times Colonist* (Victoria, BC), 22 August 1972.

"Wild Man of the Woods." *Courier-Journal* (Louisville, KY), 24 October 1878.

"Woman Battles 'It' in Ohio River." *Evansville Press* (Evansville, IN). 15 August 1955.

Index

Also from Anomalist Books...

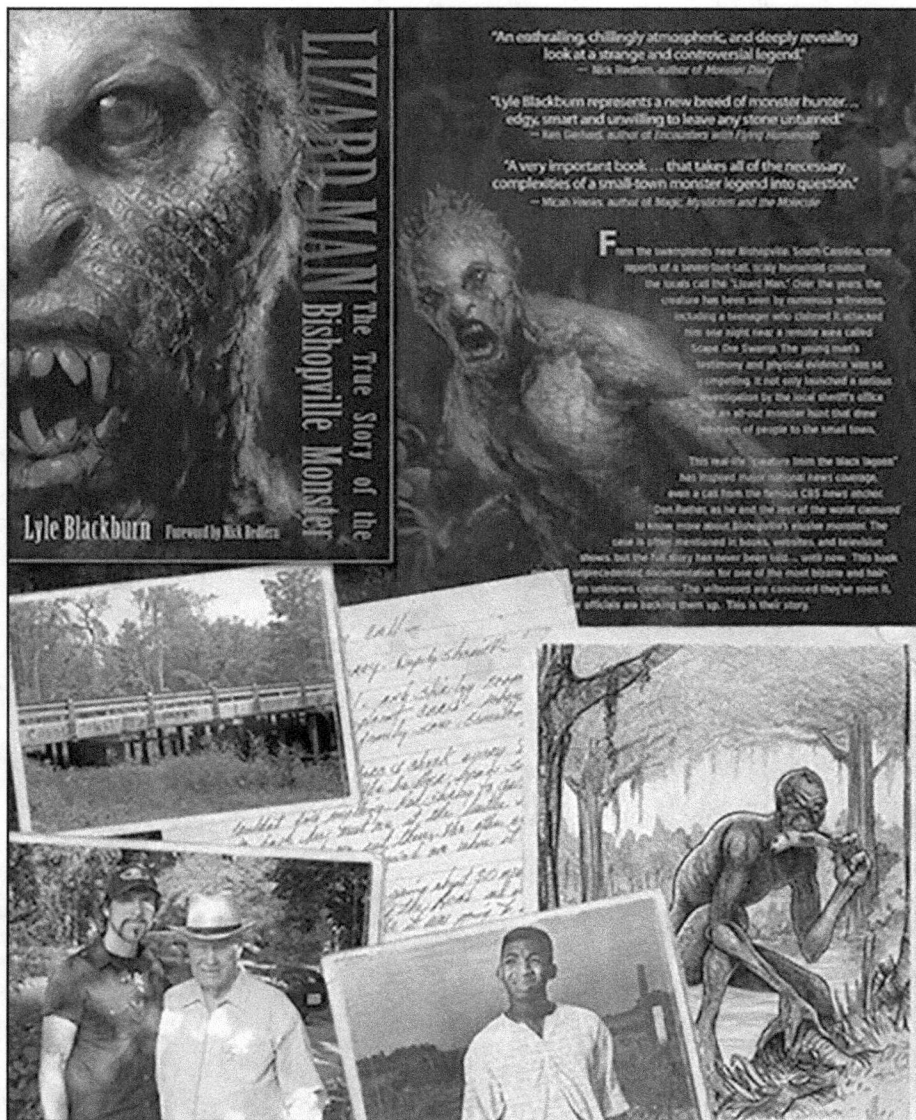

LIZARD MAN: The True Story of the Bishopville Monster
by Lyle Blackburn

Anomalist Books